THE BIG IDEA SCIENCE BOOK

The incredible concepts
that show how science
works in the world

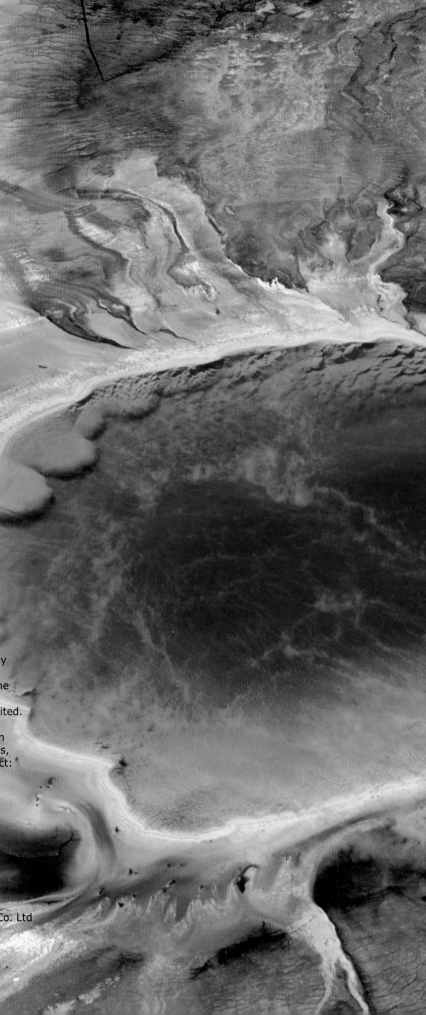

**LONDON, NEW YORK,
MELBOURNE, MUNICH, AND DELHI**

For The Book Makers Ltd:
Editorial and design Gill Denton,
Ali Scrivens, and Miranda Brown

For Dorling Kindersley:
Editor Matilda Gollon
Managing Editor Linda Esposito
Managing Art Editor Diane Thistlethwaite
Category Publisher Laura Buller
Production Editor Andy Hilliard
Production Controller Angela Graef
Jacket Designer Laura Brim
Jacket Editor Matilda Gollon
Design Development Manager
Sophia M Tampakopoulos Turner

Consultant Lisa Burke

Adapted from
The Science Reference Library, 2010
For The Book Makers Ltd:
Design Ali Scrivens and Miranda Brown
For Dorling Kindersley:
Managing Editor Sophie Mitchell
Managing Art Editor Richard Czapnik
For Pearson US:
Editorial Sharon Inglis, Stephanie Rogers,
and Eleanor McCarthy

This edition first published in the United States
in 2010 by DK Publishing
375 Hudson Street,
New York, New York 10014

13 14 10 9 8 7 6 5 4 3 2 1
001-196352-Jul/13

DK books are available at special discounts when
purchased in bulk for sales promotions, premiums,
fundraising, or educational use. For details, contact:
DK Publishing Special Markets
375 Hudson Street,
New York, New York 10014
SpecialSales@dk.com

A catalog record for this book is
available from the Library of Congress

ISBN: 978-0-7566-8902-5

Color reproduction by MDP, United Kingdom
Printed and bound in China by South China Printing Co. Ltd

**Discover more at
www.dk.com**

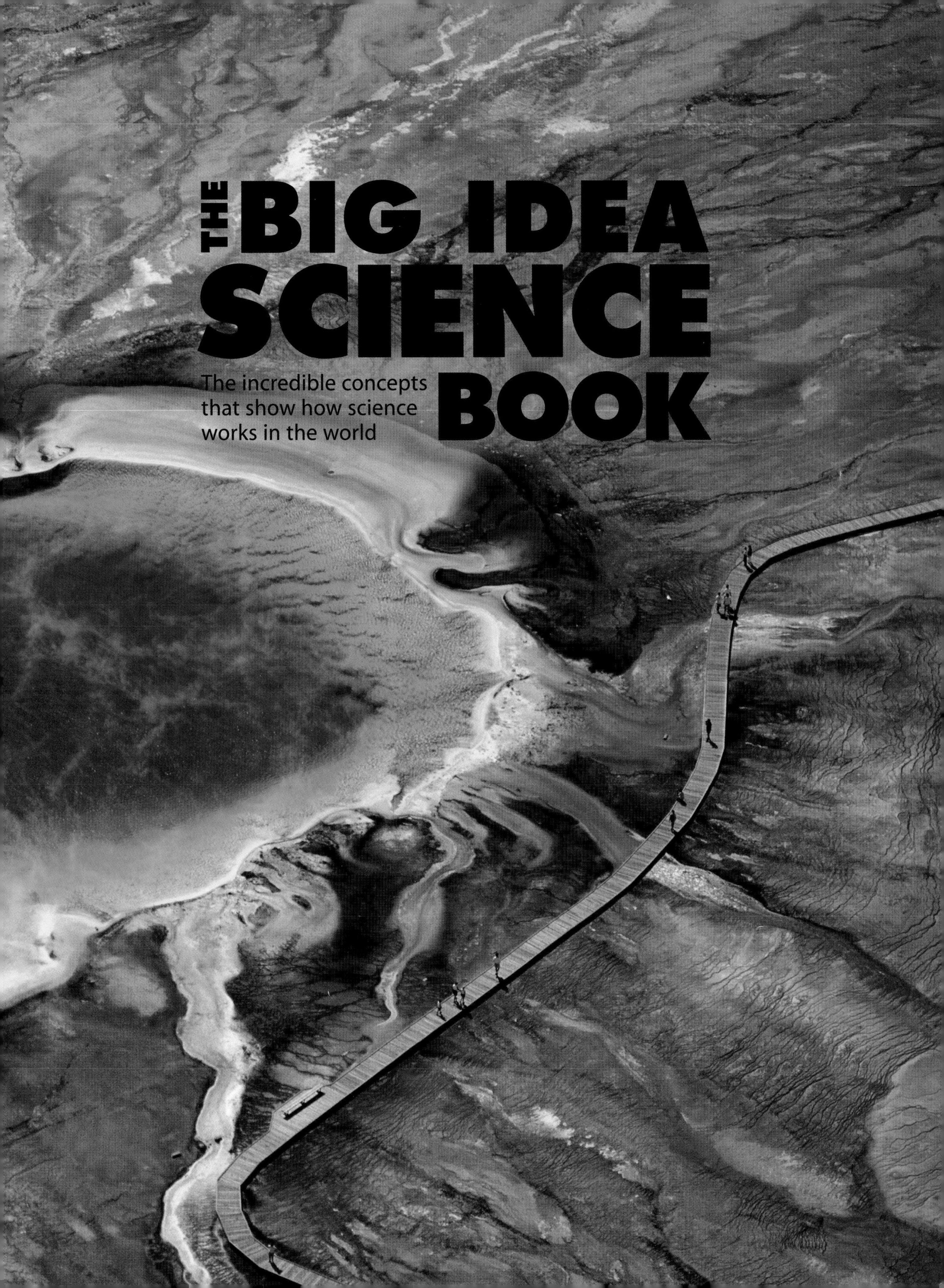

THE BIG IDEA SCIENCE BOOK

The incredible concepts
that show how science
works in the world

CONTENTS

LIFE SCIENCE

INTERACT WITH YOUR WORLD!

Watch science come alive on screen with an amazing interactive website created especially for the book. It is bursting with things to explore and do! Fantastic video clips and interactive animations take you inside plants, around the human body, deep below Earth's surface, and into the depths of space—for an even closer look at science in action!

This unique hands-on experience gives you the chance to apply everything that you have learned and see even more! Click on incredible illustrations to animate scientific processes, watch real-life science, or test your newly acquired knowledge with fun quizzes.

By interacting with science, you can really understand how it works! Seeing is learning and that's just a click away... just log on to:

http://www.children.dkonline.com

INTERACTIVE ILLUSTRATIONS

Log on and follow the simple instructions to make science spring into action!

WATCH THE MUSCLES IN MOTION

Watch a slow-motion replay of a slam dunk to see how skeletal muscles control the basketball player's movements.

 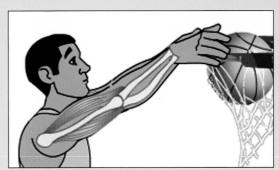

FRICTION

Follow a penguin having fun in the snow to learn about the different types of friction.

◄ Watch the penguin slide down the icy slope to learn about sliding friction.

◄ See how the penguin's moving snowball demonstrates rolling friction.

◄ Zoom in on the labeled diagram to understand the forces in action.

SEE CONTINENTAL DRIFT IN ACTION

Press play and recreate the process whereby Pangea broke up and the continents moved to where they are today.

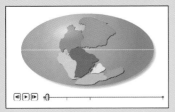 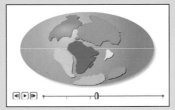 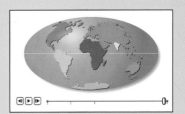

225 Million Years Ago   135 Million Years Ago   Present

Now that's active learning!

THE 24 BIG IDEAS OF SCIENCE

Science is the study of everything around us. Yet there is so much around us, how can we possibly learn everything, and where do we start? No wonder science can seem overwhelming.

Thankfully, science is not made up of discrete pieces of unrelated information that we have to learn one by one. In fact, it is built on a backbone of basic principles, which connect and help explain everything you need to know. Based on a revolutionary new approach to learning by Grant Wiggins and Jay McTighe, this book presents these key concepts as the 24 Big Ideas of Science.

Once you are familiar with these basic ideas, you will find it easier to organize information so that you don't feel flooded by random facts.

1 GENETIC INFORMATION PASSES FROM PARENTS TO OFFSPRING

2 LIVING THINGS ARE MADE OF CELLS

3 STRUCTURES IN LIVING THINGS ARE RELATED TO THEIR FUNCTIONS

4 LIVING THINGS ARE ALIKE YET DIFFERENT

5 LIVING THINGS GROW, CHANGE, AND REPRODUCE DURING THEIR LIFETIMES

6 LIVING THINGS CHANGE OVER TIME

7 LIVING THINGS GET AND USE ENERGY

8 LIVING THINGS INTERACT WITH THEIR ENVIRONMENT

9 LIVING THINGS MAINTAIN CONSTANT CONDITIONS INSIDE THEIR BODIES

10 SCIENTISTS USE SCIENTIFIC INQUIRY TO EXPLAIN THE NATURAL WORLD.

11 EARTH IS 4.6 BILLION YEARS OLD AND THE ROCK RECORD CONTAINS ITS HISTORY

12 EARTH IS THE WATER PLANET

13 EARTH IS A CONTINUALLY CHANGING PLANET

14 EARTH'S LAND, WATER, AIR, AND LIFE FORM A SYSTEM

15 HUMAN ACTIVITIES CAN CHANGE EARTH'S LAND, WATER, AIR, AND LIFE

16 SCIENCE, TECHNOLOGY AND SOCIETY AFFECT EACH OTHER

17 ATOMS ARE THE BUILDING BLOCKS OF MATTER

18 A NET FORCE CAUSES AN OBJECT'S MOTION TO CHANGE

19 ENERGY CAN TAKE DIFFERENT FORMS BUT IS ALWAYS CONSERVED

20 MASS AND ENERGY ARE CONSERVED DURING PHYSICAL AND CHEMICAL CHANGES

21 WAVES TRANSMIT ENERGY

22 THE UNIVERSE IS VERY OLD, VERY LARGE, AND CONSTANTLY CHANGING

23 EARTH IS PART OF A SYSTEM OF OBJECTS THAT ORBIT THE SUN

24 SCIENTISTS USE MATHEMATICS IN MANY WAYS

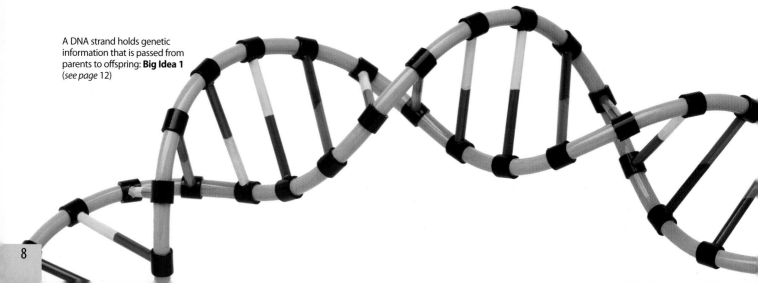

A DNA strand holds genetic information that is passed from parents to offspring: **Big Idea 1** (*see page 12*)

The book is divided into the three key areas of science: Life, Earth, and Physical. However, the Big Ideas show that there are no real boundaries in knowledge, and that by understanding a big idea in one area of science you can transfer that understanding to another seemingly unconnected subject. So, learning about human digestion will help you when you read about how a coral reef survives, because both subjects link back to Idea 3: Structures in living things are related to their function.

Life Science

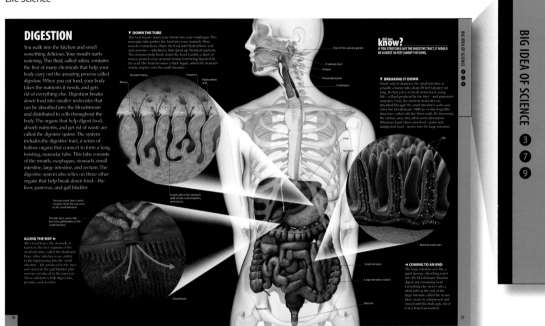

◀ You will see that each topic is clearly linked back to the **Big Ideas** by the numbers running down the side of each page. So you can easily refer back and see how different topics are connected. One subject often demonstrates a range of scientific principles.

Earth Science

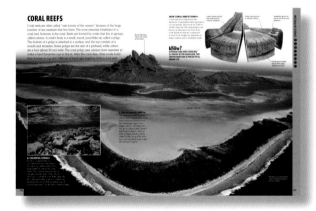

Physical Science

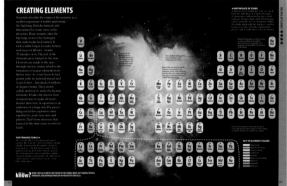

◀ Each chapter is colour coded

Why do tigers have stripes? (*see page* 71)
What makes a rainbow arc-shaped? (*see page* 176)
How does sound travel? (*see page* 222)

You can explore the Big Ideas by asking questions. Questioning is the beginning of scientific investigation. An inquisitive mind is a scientific mind. The more you know, the more you will want to know, and the more questions you will have. To keep questioning is the key to understanding our world. With this book, you will see that from a simple question, you can learn a lot, and that the 24 Big Ideas can help you to link and bring meaning to what you have learned.

LIFE SCIENCES

Life science is the study of living things, but how do we define "life"? It's not as simple as you might think. But life scientists have devised a list of characteristics that distinguish all living things: they are made of cells; maintain constant internal conditions; respond and adapt to their environment; take in and use energy; get rid of waste; grow, develop, reproduce, and pass on traits. Therefore, life science encompasses a vast array of topics, ranging from the simple cell, to cutting-edge medicine, animal behavior, GM crops, and the complexities of the human brain. As different as they might seem, all life forms, from microbes to mammals, plants to parasites, start out with a cell that holds hereditary information (DNA).

DNA CONNECTIONS

Did you realize that a fish is related to a banana tree? In fact, all living things on Earth—people, zebras, yeast, and plants—are related and share a fundamental structure of life: DNA. DNA, short for deoxyribonucleic acid, is a large molecule that carries the information an organism needs to grow and develop. Simple one-celled organisms have DNA, and multicelled organisms, such as animals, plants, and fungi, have DNA. By comparing the DNA of two different species, scientists can estimate how closely they are related. In general, closely related species have more DNA in common than distantly related species. Organisms of the same species hardly differ in their DNA at all. For example, your DNA is 99.9 percent identical to the person next to you and to all humans on Earth.

DNA COUSINS ▶

Scientists can sometimes use DNA to estimate how closely related different species are. Scientists can compare the DNA sequence—the arrangement of the DNA components—of two species. In general, the more differences there are between the sequences, the more time has passed since these two species shared a common ancestor. For instance, chimpanzees and orangutans share about 97 percent of their DNA sequence. This means that they are very closely related.

Ninety percent of DNA sequences that cause disease in humans are the same in mice, explaining their popularity in disease research.

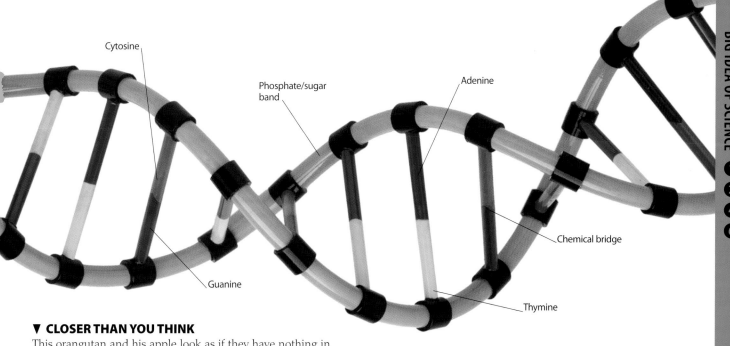

Cytosine

Phosphate/sugar band

Adenine

Chemical bridge

Guanine

Thymine

▼ CLOSER THAN YOU THINK

This orangutan and his apple look as if they have nothing in common. The apple is a plant, while the orangutan is an animal. The apple has a waxy covering, and the orangutan has skin covered in fur. But despite the differences in appearance, both the apple and the orangutan were built from instructions coded in DNA. Their DNA, and the DNA of every other living thing, is composed of the same four chemicals: A, G, C, and T. Those four chemicals are all that is needed to produce living things as different as an apple and an orangutan, bacteria and mushrooms, an oak tree and a bumblebee.

▲ STRUCTURE OF DNA

The shape of a DNA strand is like a spiraling ladder. Look at the model above. Along the sides, you can see the chain of sugar and phosphate molecules that make up the backbone of the ladder. The rungs of the ladder are formed by chemicals called *bases*. The four bases found in DNA are adenine (A), guanine (G), cytosine (C), and thymine (T). A single base sticks out from the backbone and forms a chemical bond with the base directly across from it. These two bonded bases are called *a base pair*. Adenine always pairs with thymine, and cytosine always pairs with guanine.

BLUE ZEBRA CICHLID ▲

Over time, small changes to DNA, called *mutations,* can occur. The more time that passes, the more mutations can happen. These mutations can result in new species forming. This blue zebra cichlid is one of 2,000 species of cichlids that has evolved in the last 10,000 years. That amounts to about one new species every five years—one of the fastest evolutionary waves on record.

did you know?..................
HUMANS CARRY THE DNA SEQUENCE FOR A TAIL! BUT DURING EARLY DEVELOPMENT, ANOTHER SEQUENCE OVERRIDES IT.

DNA EVIDENCE

How can scientists use genetic information to identify a criminal suspect? The answer lies in our DNA. Every person's DNA—short for deoxyribonucleic acid—is 99.9 percent the same. It is the 0.1 percent difference that can help solve crimes. Crime investigators look at 13 regions of human DNA. These areas have a great deal of variation. When DNA from a crime scene and DNA from a suspect match all 13 regions, the probability that they are from the same person is almost 100 percent. It takes only one difference in one region to prove they are not from the same person. People imprisoned before DNA evidence was available have been proven innocent and released because of that difference.

WHOSE BLOOD? ▶
An individual's DNA is the same in every cell, including blood cells. If scientists collect the DNA from blood at a crime scene, they can use the particular arrangement of molecules, called *DNA sequences*, to identify a criminal or a victim. Even if no one saw the crime, the DNA might be able to tell police who was involved.

Human skin   Human hair   Loose scales of skin around the follicle

▲ EVEN HAIR HAS DNA
Criminal cases have been solved by DNA analysis of saliva on cigarettes, stamps, cups, or mouth openings on ski masks used in a crime. Even a single hair, without the follicle, can reveal information. The DNA in hair, bones, and teeth comes from a cell's mitochondria rather than from its nucleus. The DNA that is in the mitochondria, unlike the DNA that is in the nucleus, does not contain all of the information, because it is inherited only from the mother. However, it lasts longer, so it is often used in older unsolved "cold" cases. It can be used to exclude a suspect, but not to convict one.

HAIR FOLLICLE ▲
DNA is found in the sac, called a *hair follicle*, where a hair attaches to the body, as well as in skin, bone, teeth, saliva, sweat, earwax, and even dandruff!

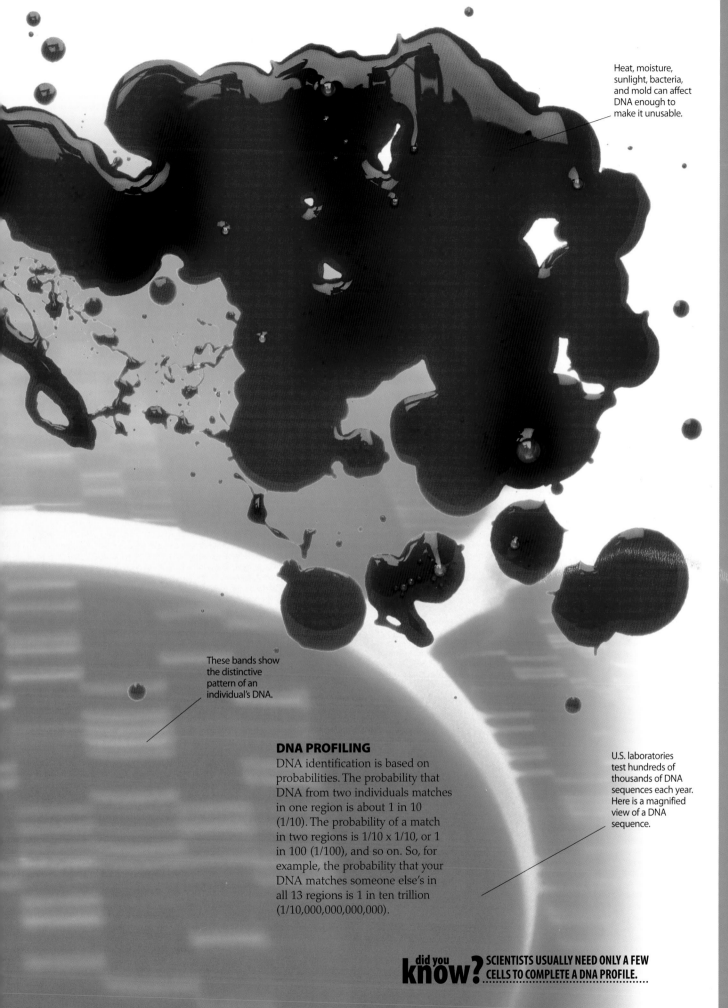

Heat, moisture, sunlight, bacteria, and mold can affect DNA enough to make it unusable.

These bands show the distinctive pattern of an individual's DNA.

DNA PROFILING

DNA identification is based on probabilities. The probability that DNA from two individuals matches in one region is about 1 in 10 (1/10). The probability of a match in two regions is 1/10 x 1/10, or 1 in 100 (1/100), and so on. So, for example, the probability that your DNA matches someone else's in all 13 regions is 1 in ten trillion (1/10,000,000,000,000).

U.S. laboratories test hundreds of thousands of DNA sequences each year. Here is a magnified view of a DNA sequence.

did you know? SCIENTISTS USUALLY NEED ONLY A FEW CELLS TO COMPLETE A DNA PROFILE. ·············

HUMAN GENOME

Scientists have put together a puzzle that has more than 3 billion pieces. The puzzle is called the *human genome,* a full set of all the genetic information in human DNA (deoxyribonucleic acid). Scientists already knew certain things about the puzzle when they began the Human Genome Project in 1989. They knew where to find DNA—in the nucleus of each human cell, on the structures called *chromosomes.* They knew what DNA looks like—a twisted ladder, with rungs made of four different chemicals, called *nitrogen bases.* They learned that DNA can be divided into 20,000 to 25,000 sections, each of which is called a *gene.* One gene might be made up of anywhere from thousands to millions of bases. To complete the puzzle, scientists had to learn the order, or sequence, of every one of the 3 billion bases. Different groups of scientists have worked on the puzzle, one finishing it in 2001 and another in 2003, and published the sequence of the basic human genome. The challenge now is to find out which human traits, structures, and diseases are influenced by which parts of our amazing genome.

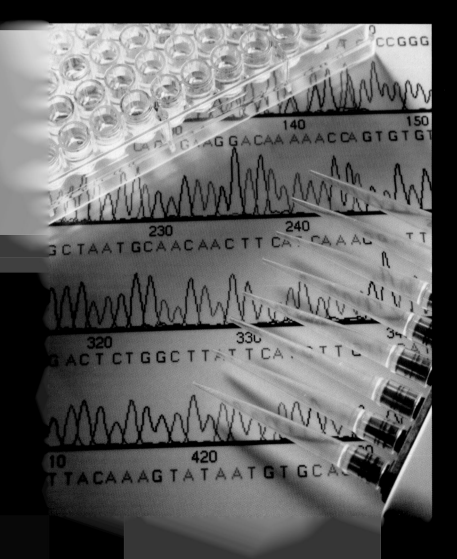

READING FRAGMENTS OF DNA ▶
DNA sequencing that used to take years is now a much faster, automated process. Multiple fragments of DNA can be analyzed at one time. The process includes many steps between extracting the DNA from a cell and analyzing its sequence of bases. Liquid containing DNA is inserted into a thick gel, and in a process called *electrophoresis,* electricity is used to sort the fragments of DNA. The gel is then viewed on a lightbox that uses ultraviolet light (shown to the right). A computer analyzes the DNA sequence, identifying the order in which the four bases occur.

◀ SPELL CAT, TAG, ACT
The four nitrogen bases in DNA are adenine, thymine, guanine, and cytosine, which are referred to as A, T, G, and C. This computer printout shows the sequence in which they occur in a fragment of DNA. Every human gene has a particular sequence of bases. Some sequences tell a cell to make a particular type of protein. Others do not code for protein, and scientists are still analyzing their purpose. Scientists are working to understand how one DNA sequence translates into a protein found in a brain tumor, while another translates into a protein found in a healthy brain cell.

did you
know?
THE LARGEST KNOWN HUMAN GENE HAS 2,400,000 BASES.
MISSING OR DUPLICATED BASES IN THIS GENE CAN CAUSE THE
MUSCLE-WEAKENING DISEASES CALLED *MUSCULAR DYSTROPHIES*.

CELL DIVISION

A person, an elephant, and a snake look very different from one another. Yet all three begin life as a single cell. So how does that cell become an adult elephant, with trillions of cells? It all starts with cell division. The first cell splits into two cells, two cells split into four, four cells split into eight, and so on. After three days, the cluster of cells, called the elephant *embryo,* consists of approximately 30 cells—called *embryonic stem cells.* These stem cells have the amazing ability to become any type of cell in the body—blood cells, brain cells, heart muscle cells, bone cells, or even hair cells in the inner ear! As the elephant's stem cells continue to divide, they become the different types of cells that together make an elephant.

Before a cell splits, it makes a copy of its genetic information, or DNA.

The nucleus of the cell splits, and the original and duplicate DNA move to opposite ends of the cell.

A cell membrane begins to form around each nucleus as the cell pulls apart.

Each new cell now has one copy of the genetic information.

▲ SPLITTING UP

Cell division helps organisms grow larger—from a single cell into a 12,000-pound (5,443-kg) adult elephant, for example. Cells also divide to repair and replace parts of the body. The cells on the edge of a cut divide to form new skin. Dead skin cells are constantly being replaced by newly divided cells. Some other adult cells, such as nerve cells, do not divide as often.

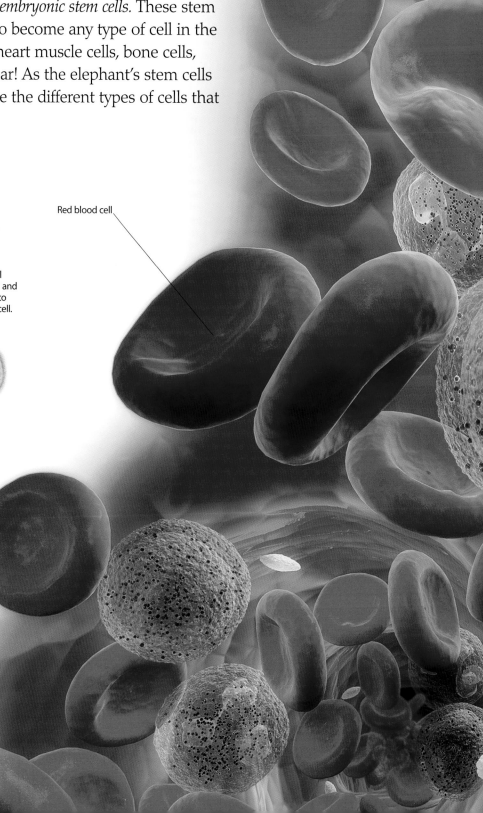

Red blood cell

◀ WHAT KIND OF CELL WILL I BE?

Once an elephant—or a person—becomes an adult, it has fewer stem cells. It does have some, though, called *adult stem cells.* In the bone marrow, for example, stem cells keep dividing to replace old cells. These stem cells can become red or white blood cells or platelets, each of which has a different job. The organism's DNA and signals throughout the body determine what type of cell each stem cell should become.

A lymphocyte, a type of white blood cell, targets infections and cancers.

did you know?
STEM CELLS IN AN AVERAGE ADULT'S BONE MARROW GENERATE ABOUT 610 BILLION BLOOD CELLS PER DAY!

Inner hair cells transmit signals to the brain.

Outer hair cells receive vibrations.

Nerve fiber

This platelet will clump with other platelets to help blood clot when we cut ourselves.

White blood cells destroy harmful foreign organisms. This neutrophil, the most common type of white blood cell, targets harmful bacteria.

HAIR FOR HEARING ▲

Inside a mammal's inner ear is a chamber, called the *cochlea,* where sensory cells, called *hair cells* because of their tiny hairlike projections, help transmit sound. Damage to these hairs causes hearing loss. Researchers are exploring ways to grow stem cells that may generate new hair cells.

MUTATIONS

Why do some people have brown hair and some people have red hair? The simple answer is genes. Genes, regions of a person's chromosomes, direct cells to produce specific proteins. These proteins help determine the physical traits of a person or any other living thing. But even though cells and cellular processes are pretty amazing, they are not always perfect. Sometimes a change in the DNA of a gene, called a *mutation*, can occur and cause a cell to make an incorrect protein. Since proteins affect an organism's physical traits, mutations in the genes that make these proteins can alter an organism's traits. Red hair, with its accompanying freckles and light-colored skin, is a mutation. So is a genetic disorder such as Type 1 diabetes. Mutations can be helpful, harmful, or neither. Mutations contribute to the astonishing diversity of living things.

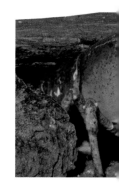

◄ FIVE-LEGGED SHEEP

Although it seems rare, there have been cases all over the world of animals born with extra limbs. The mutation of a gene involved in limb development can cause extra limbs to form. Depending on the situation, many of these animals can live happily. This five-legged sheep was born in 2002 in the Netherlands. Her owner said she was able to live with her extra limb without problems. A lamb in New Zealand was born with seven legs. It unfortunately was unable to survive because of other health issues.

WHITE TIGER ►

White tigers can be born when both parents carry a recessive gene for the white color. The majority of white tigers are found in captivity. They are at a disadvantage in the wild and, therefore, are very rare there. Orange and black tigers can hide in the jungle. It's more difficult than you would think to spot a tiger among jungle plants. But a white tiger is much more visible, making hunting without being seen difficult.

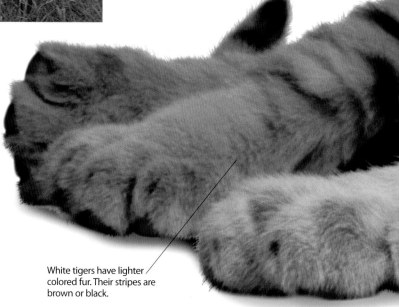

White tigers have lighter colored fur. Their stripes are brown or black.

did you know?
....
MANY ZOOLOGISTS BELIEVE THAT ALL WHITE TIGERS IN THE UNITED STATES ARE THE DESCENDANTS OF A SINGLE WHITE TIGER.

White tigers usually have blue eyes, while typical tigers have yellow eyes.

BLUE LOBSTER ▲

If you could pick what color lobster you'd like to be, you might want to choose blue. A blue lobster's color is the result of a mutation that causes excess production of a certain protein. These lobsters are rare, and when they're caught, they most often end up in zoos and aquariums instead of a cooking pot. In this case, the mutation is definitely a good thing.

FRANKENFOODS

The fictional character, Victor Frankenstein, was obsessed with creating life. He used old body parts to build a creature. After he brought the creature to life, he was horrified by what he had made—a monster. Should people create new types of food crops, or is there a danger of creating "Frankenfoods"? Opponents of altering the genetic material of food crops use this nickname for genetically modified organisms, called *GMOs* or *transgenic* crops. They point out that GMOs may have unanticipated, harmful characteristics and effects. However, GMO supporters argue that transgenic crops can have positive characteristics, such as resistance to insects or higher vitamin content. Farmers long ago figured out how to selectively breed plants, called *hybrids*, that have the best characteristics of the parent plants. GMOs, on the other hand, are created by inserting the genetic material of one individual into that of another. There is a great deal of debate over the pros and cons of GMOs. Many questions remain about their safety for humans, their effect on unmodified crops, and the rules that will govern their use.

RICE ▼

Billions of people in Asia depend on rice as their main source of calories. Some rice now on the market has been genetically modified to contain more vitamin A (beta carotene), iron, and zinc. Vitamin A deficiency can cause malnutrition and blindness. One type of rice was developed using genes from daffodils and bacteria. Is it safe to eat this rice? In the short term, it appears that GMOs are safe. However, people have not been eating GMOs long enough for us to know whether there are any long-term effects.

did you know?

ABOUT 80 PERCENT OF THE CORN PLANTED IN THE UNITED STATES IN 2008 WAS FROM GENETICALLY MODIFIED SEED.

CORN ▼

Genes used to create GMOs may come from different types of organisms. For example, some insect-resistant corn has genetic material from a type of bacteria. Pollen from this corn has blown over the U.S. border or been planted by farmers in Mexico, where planting most GMO corn is banned. GMO opponents do not want this altered corn to breed with the native varieties of corn that grow in Mexico.

Although commercially grown strawberries are larger than these wild Alpine ones, they typically do not have their intense flavor.

The corn earworm is the most serious sweet-corn pest, feeding directly on corn kernels.

STRAWBERRIES ▲

Many opponents of GMO foods point out that plants can be bred to have certain traits over time, using traditional selective breeding techniques. Sometimes this happens naturally. Commercial strawberries that we eat today are a hybrid of two different strawberry plants that bred accidentally in Europe in the mid-1700s. These strawberries, larger than those of the parent plants, are now raised all over the world.

TOMATOES ▲

The first genetically modified tomatoes came onto the market in 1994. They were engineered so that they did not produce an enzyme that caused them to rot. This modification helped them stay fresh longer. However, they also contained genes that made them resistant to antibiotics. After doctors voiced concern that these genes could be transferred to bacteria in the human gut, these tomatoes were taken off the market.

When a cotton boll is mature, it bursts open to show the fluffy white seed fibers.

◄ SOYBEANS

Nearly all soybeans produced in the United States come from genetically modified seeds. They are designed to be resistant to herbicides that are used to kill weeds. However, in 2009, more farmers began planting non-genetically modified soybeans again because the price of GMO seeds had become too high.

COTTON ▲

Cotton has been genetically modified to resist pests. The bollworm is an insect that can do extensive damage to cotton crops.

BRAIN POWER

For nearly everything you do, a part of your brain is in charge. The brainstem controls your most basic functions—heartbeat, breathing, digestion. The brainstem also relays messages into and out of other parts of your brain. When you feel the warmth of a campfire, the sensation travels through nerves from your skin to your spinal cord and into your brainstem. You move toward the fire—the command to move went from your cerebellum through your brainstem and out to your muscles. You reach out to the fire—ouch! But by the time the pain signal gets to your thalamus and it tells the gray matter of your brain that your hand is getting too warm, your hand is already pulling away. That's because a few messages are too urgent to wait. As soon as the news "Too hot to handle!" reaches your spinal cord, "Get out, NOW!" starts back toward your hand. Pulling back is what's called a *reflex action,* which travels from your hand to your spinal cord and back to your hand, without going through your brain. And saying "ouch"? Thank the speech area of your brain.

RIGHT IS LEFT, LEFT IS RIGHT ▼

The brain is divided into two halves, called *hemispheres.* Although the two sides of the brain look symmetrical—the same on both sides—they handle different tasks. While you read these words, your left brain works more. Or at least that's the case for almost all right-handed people. About 40 percent of left-handed people use either the right hemisphere or the whole brain for language. The right brain appears to be more involved in visual recognition of people and objects.

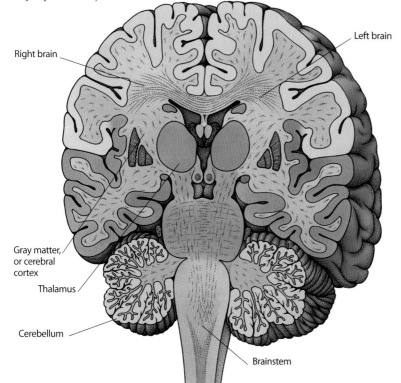

Right brain

Left brain

Gray matter, or cerebral cortex

Thalamus

Cerebellum

Brainstem

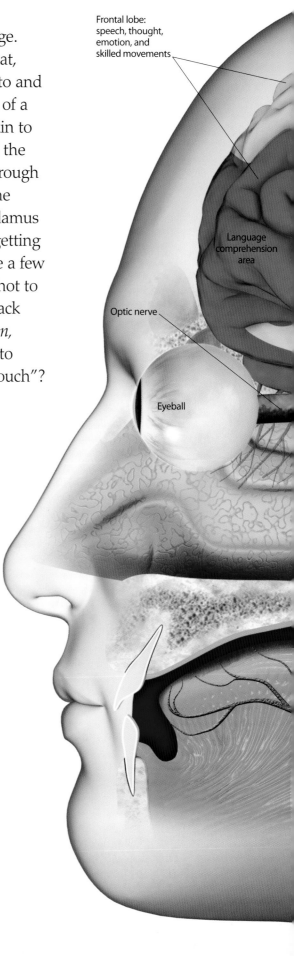

Frontal lobe: speech, thought, emotion, and skilled movements

Language comprehension area

Optic nerve

Eyeball

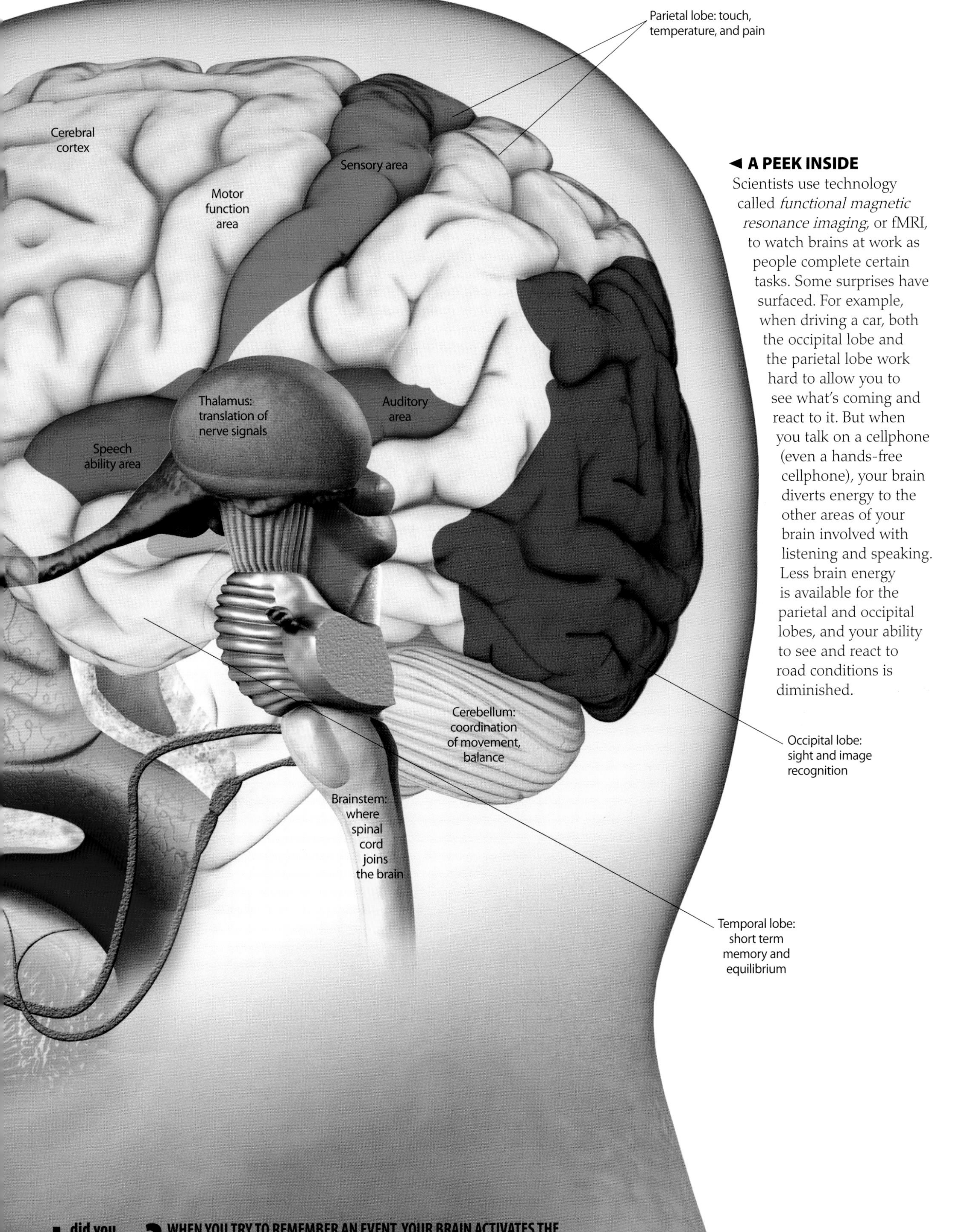

Cerebral
cortex

Motor
function
area

Sensory area

Parietal lobe: touch,
temperature, and pain

Thalamus:
translation of
nerve signals

Auditory
area

Speech
ability area

Cerebellum:
coordination
of movement,
balance

Brainstem:
where
spinal
cord
joins
the brain

Occipital lobe:
sight and image
recognition

Temporal lobe:
short term
memory and
equilibrium

◄ **A PEEK INSIDE**

Scientists use technology
called *functional magnetic
resonance imaging,* or fMRI,
to watch brains at work as
people complete certain
tasks. Some surprises have
surfaced. For example,
when driving a car, both
the occipital lobe and
the parietal lobe work
hard to allow you to
see what's coming and
react to it. But when
you talk on a cellphone
(even a hands-free
cellphone), your brain
diverts energy to the
other areas of your
brain involved with
listening and speaking.
Less brain energy
is available for the
parietal and occipital
lobes, and your ability
to see and react to
road conditions is
diminished.

did you know? WHEN YOU TRY TO REMEMBER AN EVENT, YOUR BRAIN ACTIVATES THE
SAME AREAS OF THE BRAIN THAT WERE ACTIVE DURING THE EVENT.

LEFT VS. RIGHT BRAIN

Your cerebrum is made up of a left and a right hemisphere. The two hemispheres are connected by a bundle of nerve fibers. The two sides work together to control just about everything you do. Research about the particular capabilities of each side of the brain constantly yields new information. We know that movement of one side of the body is generally controlled by the opposite hemisphere. For example, the left brain controls the right hand. For most people, the right hand is dominant, so their left hemisphere is sometimes considered dominant. The dominant hemisphere is also the usual location for processing language. Almost all right-handed people process language in the left brain. But 60 percent of left-handed and ambidextrous (can use both hands equally well) people also process language in the left brain, with the rest processing in the right brain or in both hemispheres. The right brain appears to be more involved in processing spatial information and recognizing faces.

STUDY SKILLS ▲
Different study skills tend to use different hemispheres of your brain. Making lists and classifying are considered left brain tasks. Stepping back to see "the big picture" is considered a right brain function.

CALCULATIONS ▲
Scientists think we use the left side of our brains more when we try to solve mathematical equations. The left hemisphere seems to be dominant for math and logic.

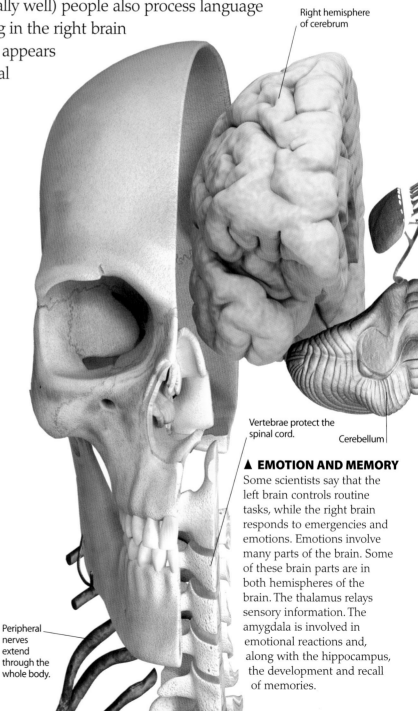

Right hemisphere of cerebrum

Vertebrae protect the spinal cord.

Cerebellum

Peripheral nerves extend through the whole body.

▲ EMOTION AND MEMORY
Some scientists say that the left brain controls routine tasks, while the right brain responds to emergencies and emotions. Emotions involve many parts of the brain. Some of these brain parts are in both hemispheres of the brain. The thalamus relays sensory information. The amygdala is involved in emotional reactions and, along with the hippocampus, the development and recall of memories.

DRAWING A FACE ▲
The right hemisphere of this artist's brain may help in both visualizing the face to be drawn and focusing on developing the many parts of the picture at once.

◄ **HEARING MUSIC**
Music activates both sides of the brain. The left brain seems to process rapid changes in frequency and intensity. The right side seems to perceive pitch and melody.

did you know?
MUCH OF WHAT WE KNOW ABOUT THE BRAIN HEMISPHERES HAS COME FROM PEOPLE WHOSE SEVERE SEIZURES WERE STOPPED BY SURGICALLY REMOVING THE CONNECTION BETWEEN THE TWO HALVES OF THEIR BRAINS.

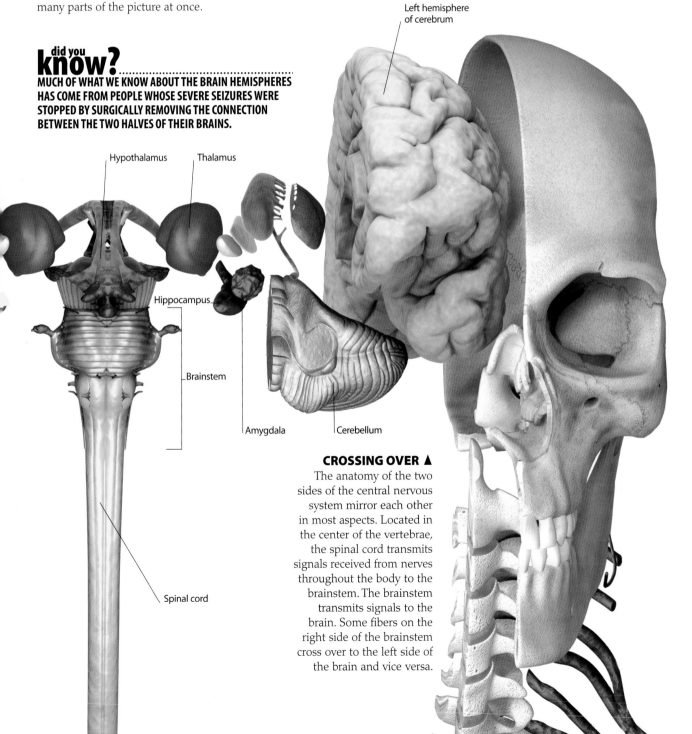

Hypothalamus   Thalamus

Hippocampus

Brainstem

Amygdala   Cerebellum

Spinal cord

Left hemisphere of cerebrum

CROSSING OVER ▲
The anatomy of the two sides of the central nervous system mirror each other in most aspects. Located in the center of the vertebrae, the spinal cord transmits signals received from nerves throughout the body to the brainstem. The brainstem transmits signals to the brain. Some fibers on the right side of the brainstem cross over to the left side of the brain and vice versa.

HYPOTHALAMUS

Near the base of your brain lies a group of specialized cells called the *hypothalamus*. It controls the autonomic nervous system, which regulates breathing, blood pressure, and heart rate. The hypothalamus also releases chemicals that travel to the pituitary gland to stimulate or suppress the release of hormones. Hormones are chemical messengers that regulate and coordinate processes in the body. Pituitary hormones influence growth, sexual development, and metabolism. Parts of the hypothalamus regulate blood sugar levels, sleep cycles, thirst, hunger, 24-hour rhythms, energy levels, and emotions. The hypothalamus also controls body temperature. When the body temperature is too high or too low, the hypothalamus sends out signals to adjust the temperature. If you are too hot, for example, the hypothalamus sends signals that cause the capillaries in your body to expand. Expanded capillaries help your blood cool itself faster and can make your face look flushed.

THE PITUITARY GLAND ▼

Just below the hypothalamus is the pituitary gland, a pea-size gland that controls hormone production. The release of pituitary hormones can be influenced by your emotions or by changes in the season. The hypothalamus senses environmental temperature, daylight patterns, and feelings. It sends this information to the pituitary, which may cause the pituitary to release more or fewer hormones. The pituitary can produce chemicals that elevate mood and reduce feelings of pain.

did you know? ....
IN SPITE OF ALL IT DOES, THE HYPOTHALAMUS IS ABOUT THE SIZE OF AN ALMOND!

THE ENDOCRINE SYSTEM

The endocrine system works with the nervous system to keep the body functioning properly. The body's glands and hormones form the foundation of the endocrine system, which also includes the pituitary, thyroid, parathyroid, pineal, and adrenal glands, and the gonads. The hypothalamus links the two systems. Problems in the endocrine system can lead to diseases and disorders such as diabetes, osteoporosis (decreased bone mass), and growth and development problems.

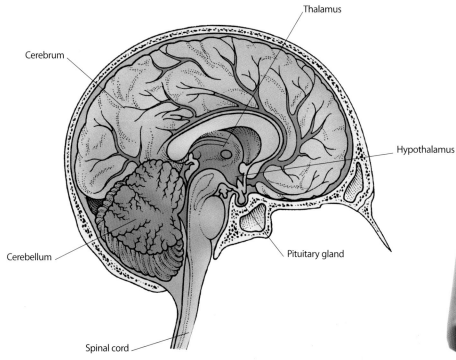

Thalamus

Cerebrum

Hypothalamus

Cerebellum

Pituitary gland

Spinal cord

Bao Xishun, one of the tallest men on record, grew normally until he was 16. Large growth spurts brought him to his present height of 7 feet 8.95 inches (2.36 m) by age 23.

GROWTH SPURT ▲

Puberty-related body development occurs in girls between ages 10 and 14 and boys between ages 12 and 16. The boys above are all 13 or 14 years old. The hypothalamus signals the pituitary gland to release growth hormones. These hormones control sexual development, growth spurts, 24-hour rhythms, and the menstrual cycle.

He Pingping, the shortest man on record at 2 feet 5.37 inches (74.59 cm), suffers from osteogenesis imperfecta, a genetic disorder that causes brittle bones and short stature, among other symptoms.

ALS

Patients with ALS, or amyotrophic lateral sclerosis, may feel like prisoners in their own bodies. Yet, with help, some survive for decades to live very full lives. ALS is a disease that results from the gradual weakening of the nerves, called *motor neurons*, that control muscle movement. When the nerves stop working, the muscles they control get weaker and thinner. People with ALS slowly lose control of their arms, legs, and even the muscles that allow them to speak. They eventually become paralyzed. Thankfully, they usually keep their ability to think, sense, or understand the world around them. Scientists do not know what causes ALS. Some forms of the disease can be inherited. Viruses or environmental toxins may play a role as well. ALS is also known as Lou Gehrig's disease, after a famous American baseball player who had the disease.

LIFE WITH ALS ▲

Stephen Hawking, a famous British theoretical physicist, has lived with ALS for over 40 years. He was diagnosed when he was 21, after noticing that he had become clumsy. Over time, he became dependent on a wheelchair and had to use a speech synthesizer to speak. With help from his wife, children, nurses, assistants, and technology, he has written important research papers and best-selling books, given many lectures, appeared in movies and television, and has even been in space.

did you know?
MOTOR NEURONS CAN BE MORE THAN 3 FEET LONG. THE LONGEST ONE RUNS FROM YOUR LOWER SPINE TO YOUR BIG TOE.

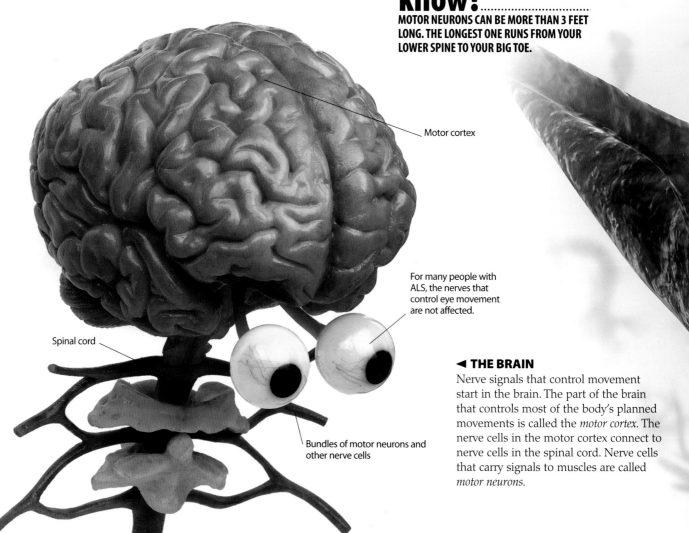

Motor cortex

For many people with ALS, the nerves that control eye movement are not affected.

Spinal cord

Bundles of motor neurons and other nerve cells

◄ THE BRAIN

Nerve signals that control movement start in the brain. The part of the brain that controls most of the body's planned movements is called the *motor cortex*. The nerve cells in the motor cortex connect to nerve cells in the spinal cord. Nerve cells that carry signals to muscles are called *motor neurons*.

▼ MOTOR NEURON

A motor neuron is made up of a cell body, several short arms called *dendrites,* and one long arm called an *axon.* The cell body and dendrites receive incoming signals. The axon sends out signals. A signal is a moving electrical charge that travels from cell to cell.

Direction of signal

❶ AXON

An axon carries a signal away from a neuron. In this model, the arms with knobby ends—such as the one numbered *1*—are the tips of axons from other neurons, carrying incoming signals to this motor neuron. The signal is traveling in the direction of the arrows.

❷ SYNAPSE

The place where the knobby end passes a signal to this motor neuron is called a *synapse.* The signal has to cross the gap at the synapse to get from one neuron to another.

❸ CELL BODY

The thickest part of the cell is called the *cell body,* where the nucleus is located. From there, the signal travels out on this neuron's axon to move a muscle.

❹ MYELIN

Flat cells wrap around axons the way a bandage is wrapped around a finger. They form a material called *myelin.* Myelin acts as the insulation in an electrical wire does, keeping nerve signals strong over long distances.

❺ DENDRITE

The spokelike dendrites deliver incoming signals from other nerve cells.

SKELETONS

What words would you use to describe the word *bone*? Strong? Solid? Dead? If you said strong, you are right. Bones contain a fibrous protein called *collagen* that is combined with minerals such as calcium and phosphate. This nonliving material makes bones very strong and flexible. Solid? No. Most bones are made up of two types of bone: compact bone and spongy bone. While compact bone is very dense and firm, spongy bone is filled with small spaces that contain bone marrow. Dead? Definitely not. Most adult animals' bones contain a small percentage of living cells. Osteogenic cells, or bone stem cells, produce osteoblasts and osteoclasts. Osteoblasts make new bone tissue and eventually become osteocytes, which keep bone tissue healthy. Osteoclasts destroy damaged bone tissue, a necessary part of the bone repair process. Together, these cells allow bones to grow and to heal after injuries.

HUMAN SKELTON ▶

The adult human skeleton is made up of 206 bones. The longest is the femur, or thigh bone. It is about one quarter of a person's height. The smallest, which is the stapes in the middle ear, is only a tenth of an inch (about 2.5 mm) long. However, when you are born, you have almost 300 bones. Some are made of cartilage, a strong, flexible tissue. Others are partly made of cartilage. Over time, these grow together and become the 206 bones in an adult.

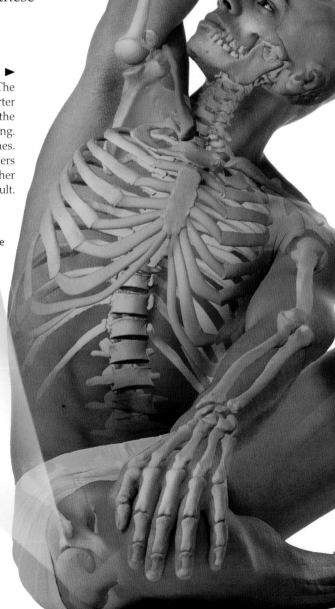

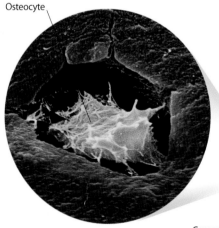

Osteocyte

Spongy bone

Compact bone

Femur

COMPACT BONE TISSUE ▲

Compact bone is made up of densely packed nonliving material that runs the length of the bone. Osteocytes are the cells that maintain the nonliving bone material by recycling calcium salts and assisting in repairs. They are located in small holes in the compact bone.

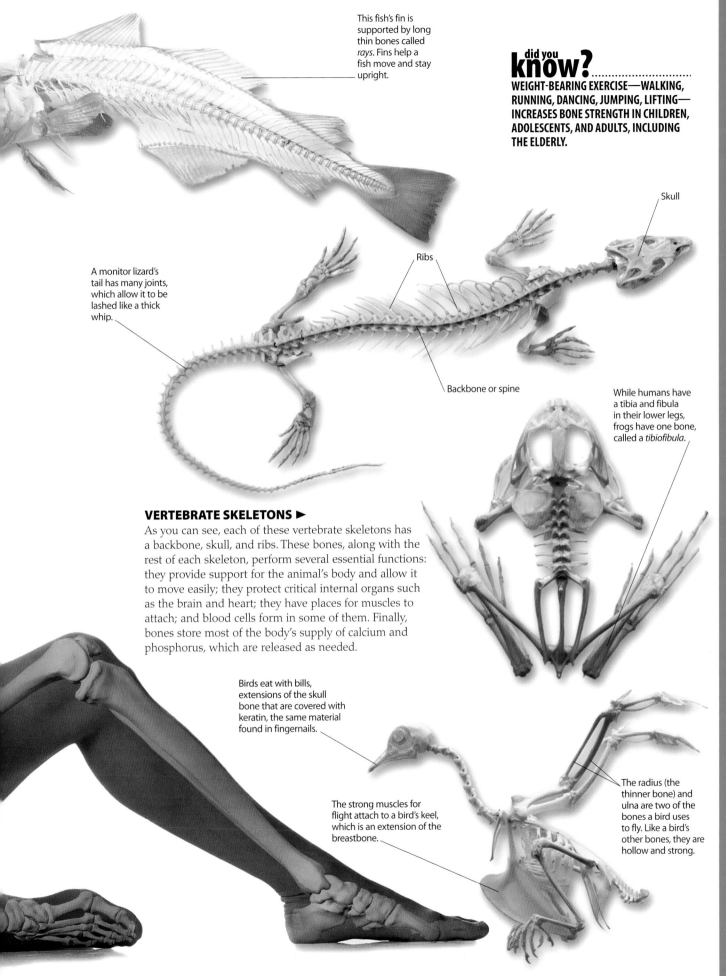

This fish's fin is supported by long thin bones called *rays*. Fins help a fish move and stay upright.

did you know?
WEIGHT-BEARING EXERCISE—WALKING, RUNNING, DANCING, JUMPING, LIFTING—INCREASES BONE STRENGTH IN CHILDREN, ADOLESCENTS, AND ADULTS, INCLUDING THE ELDERLY.

Skull

Ribs

A monitor lizard's tail has many joints, which allow it to be lashed like a thick whip.

Backbone or spine

While humans have a tibia and fibula in their lower legs, frogs have one bone, called a *tibiofibula*.

VERTEBRATE SKELETONS ▶
As you can see, each of these vertebrate skeletons has a backbone, skull, and ribs. These bones, along with the rest of each skeleton, perform several essential functions: they provide support for the animal's body and allow it to move easily; they protect critical internal organs such as the brain and heart; they have places for muscles to attach; and blood cells form in some of them. Finally, bones store most of the body's supply of calcium and phosphorus, which are released as needed.

Birds eat with bills, extensions of the skull bone that are covered with keratin, the same material found in fingernails.

The strong muscles for flight attach to a bird's keel, which is an extension of the breastbone.

The radius (the thinner bone) and ulna are two of the bones a bird uses to fly. Like a bird's other bones, they are hollow and strong.

BLOOD TYPES

Bags of blood? These bags may seem like props for a horror movie, but they actually save lives. Every two and a half seconds, someone in the world donates a pint (0.5 L) of blood. During donation, a nurse sticks a needle into a vein in the donor's arm, sending blood through a tube and into a bag. The blood is tested, and if it is free of disease, it becomes part of a blood bank. It may save a premature baby or a car crash victim. If someone needs blood, a bag of blood is connected through a tube into a patient's vein. The donor blood flows into the patient; this is called a *transfusion*. The donor and patient are strangers, but they have one thing in common: their blood type.

Donation: Blood donation is safe. The average adult body contains about 10 pints (5 L) of blood. A healthy donor's body will replace the blood cells lost from a donation within weeks.

Donation: People can receive transfusions only of human blood.

Parts: Blood banks separate donor whole blood into its parts: red blood cells, white blood cells, plasma, and platelets. Patients usually need only a single blood component instead of whole blood.

Parts: About 45% of whole blood is made up of red blood cells, which carry oxygen. If you do not have enough healthy red blood cells, you may feel tired due to a lack of oxygen.

Parts: Less than 1% of blood is made up of white blood cells, which attack germs. White blood cells float in plasma and can race to wherever they are needed.

Parts: About 55% of blood is plasma, a yellow liquid containing mostly water. Dissolved in the plasma are vitamins, hormones, and some minerals.

did you know?

A RED BLOOD CELL TRAVELS OVER 300 MILES (480 KM) THROUGH BLOOD VESSELS BEFORE IT DIES.

Antigens: Antigens are chemical molecules that sit on the surface of red blood cells. Two antigens, A and B, are the basis for the commonly used ABO system of blood typing.

Antigens: Type A blood has A antigens on its red blood cells; type B blood has B antigens; type AB blood has both; and type O blood has neither.

Antigens: The antigens in blood will make antibodies, which attack a different antigen. Type A antibodies will attack type B blood, and type B antibodies will attack type A blood.

Parts: Platelets also float around in your plasma. Platelets are sticky and help clot blood. Without platelets, you could bleed to death.

Types: People have different color eyes and hair. These are traits you can see. People also have many genetic characteristics you cannot see, such as blood type. You can have blood type A, B, AB, or O.

Types: If one of your parents has type A or B and the other has type O, you could be either type A, B, or O.

HEARTBEAT

Did you ever wonder what makes your heart beat? Like a pump in a machine, the heart squeezes and relaxes based on the careful timing of electrical signals. As the upper chambers of the heart—the atria—fill with blood, a mass of tissue called the *sinoatrial node* (sometimes called the "natural pacemaker") in the upper right part of the heart sends electrical signals to the heart muscle in the atria to tighten, or contract, and then relax. Once the atria contract, the electrical signal travels to a second node called the *atrioventricular node.* This tissue transmits the signal farther. The muscles of the heart's lower chambers—the ventricles—are then signaled to contract and then relax. This timed series of contracting and relaxing of heart muscle pumps blood through the circulatory system.

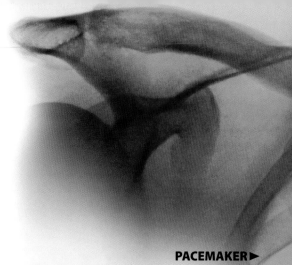

PACEMAKER ▶

If the heart's natural pacemaker, the sinoatrial node, is not working properly, the heart may beat too fast or too slow. This condition, called *arrhythmia*, can be treated with an electronic pacemaker. This X-ray shows a pacemaker that has been surgically implanted. The pacemaker can sense when the heart is beating irregularly. If that happens, the pacemaker generates an electrical signal that returns the heart to a healthy rhythm.

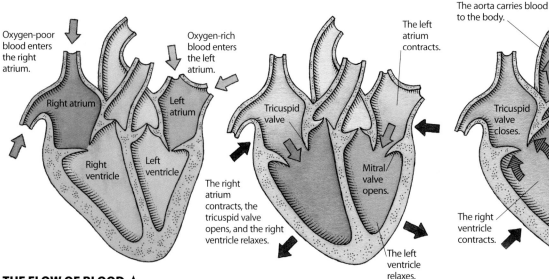

Oxygen-poor blood enters the right atrium.

Oxygen-rich blood enters the left atrium.

Right atrium

Left atrium

Right ventricle

Left ventricle

The right atrium contracts, the tricuspid valve opens, and the right ventricle relaxes.

The left atrium contracts.

Tricuspid valve

Mitral valve opens.

The left ventricle relaxes.

The aorta carries blood to the body.

The pulmonary artery carries blood to the lungs.

The pulmonary valve opens.

Tricuspid valve closes.

The aortic valve opens.

The right ventricle contracts.

The mitral valve closes.

The left ventricle contracts.

THE FLOW OF BLOOD ▲

When the atria relax, they fill with blood from the body. As they contract, the blood is pushed into the ventricles, which dilate, or get bigger. The ventricles then contract, pushing blood back to the body and lungs. Valves open and close to keep blood flowing in one direction. The cycle repeats again and again. The opening and closing of the valves also make the familiar, rhythmic sound of the heartbeat: lub-dub, lub-dub, lub-dub . . .

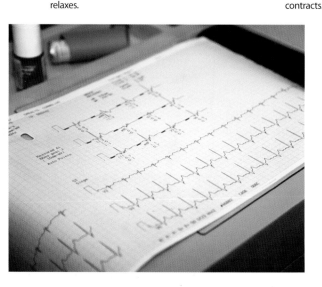

RECORDING THE HEARTBEAT ▶

A test called an *electrocardiogram,* or ECG, can tell if the heartbeat is normal. During an ECG, wires with sensors called *electrodes* are taped to the chest, arms, and legs. They sense the heart's electrical signals. These signals are recorded on graph paper. Doctors read the printout to see if the heartbeat is strong and regular.

did you
know?..........................

AN ADULT HUMAN HEART PUMPS ABOUT 100,000 TIMES PER DAY. HEALTHY NEWBORNS CAN HAVE MORE THAN TWICE AS MANY HEARTBEATS IN A DAY.

A pacemaker is connected to the heart by one or more wires, depending on the type of arrhythmia a person has.

Heart

Wire to the atrium

Wire to the ventricle

DIGESTION

You walk into the kitchen and smell something delicious. Your mouth starts watering. This fluid, called *saliva,* contains the first of many chemicals that help your body carry out the amazing process called *digestion.* When you eat food, your body takes the nutrients it needs, and gets rid of everything else. Digestion breaks down food into smaller molecules that can be absorbed into the bloodstream and distributed to cells throughout the body. The organs that help digest food, absorb nutrients, and get rid of waste are called the *digestive system.* The system includes the digestive tract, a series of hollow organs that connect to form a long, twisting, muscular tube. This tube consists of the mouth, esophagus, stomach, small intestine, large intestine, and rectum. The digestive system also relies on three other organs that help break down food—the liver, pancreas, and gall bladder.

▼ DOWN THE TUBE

The food travels down your throat into your esophagus. This muscular tube pushes the food into your stomach. Here, muscle contractions churn the food with hydrochloric acid and enzymes—substances that speed up chemical reactions. The enzymes help break down the food. Luckily, a layer of mucus protects your stomach lining from being digested by the acid. The food becomes a thick liquid, which the stomach slowly empties into the small intestine.

Stomach lining

Enzymes

Mucus

Hydrochloric acid

Gastric pits in the stomach wall secrete acid, enzymes, and mucus.

The pancreatic duct carries enzymes from the pancreas to the small intestine.

The bile duct carries bile from the gall bladder to the small intestine.

ALONG THE WAY ▶

After food leaves the stomach, it travels to the first segment of the small intestine, called the *duodenum.* Here, other substances are added to the liquid going into the small intestine—bile produced in the liver and stored in the gall bladder, plus enzymes produced in the pancreas. These substances help digest fats, proteins, and starches.

Duodenum

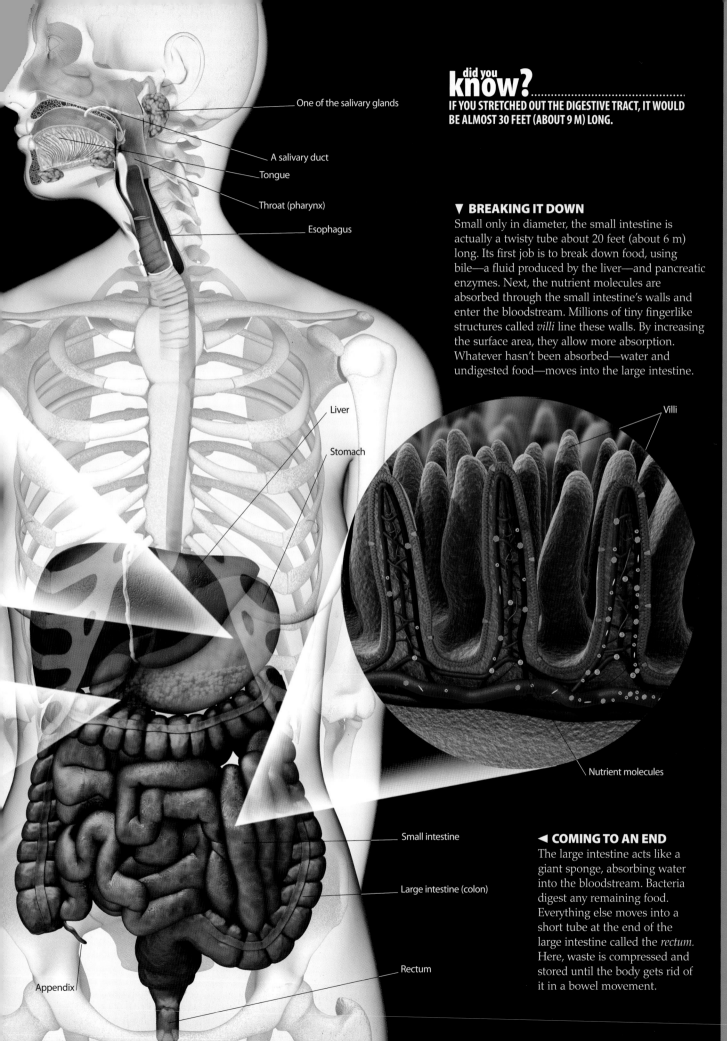

One of the salivary glands

A salivary duct

Tongue

Throat (pharynx)

Esophagus

Liver

Stomach

Small intestine

Large intestine (colon)

Rectum

Appendix

did you know?
IF YOU STRETCHED OUT THE DIGESTIVE TRACT, IT WOULD BE ALMOST 30 FEET (ABOUT 9 M) LONG.

▼ BREAKING IT DOWN
Small only in diameter, the small intestine is actually a twisty tube about 20 feet (about 6 m) long. Its first job is to break down food, using bile—a fluid produced by the liver—and pancreatic enzymes. Next, the nutrient molecules are absorbed through the small intestine's walls and enter the bloodstream. Millions of tiny fingerlike structures called *villi* line these walls. By increasing the surface area, they allow more absorption. Whatever hasn't been absorbed—water and undigested food—moves into the large intestine.

Villi

Nutrient molecules

◄ COMING TO AN END
The large intestine acts like a giant sponge, absorbing water into the bloodstream. Bacteria digest any remaining food. Everything else moves into a short tube at the end of the large intestine called the *rectum*. Here, waste is compressed and stored until the body gets rid of it in a bowel movement.

KIDNEY TRANSPLANT

Kidneys remove wastes from the blood. Without properly functioning kidneys, a body will be sickened by its own wastes. One alternative is a treatment called *dialysis*. Dialysis keeps the body in balance and does many of the tasks of the kidneys. Dialysis treatment can take up to four hours a day, three times a week, and usually is needed for the rest of a person's life. For one method of dialysis, the person is connected to an external machine that filters the wastes and excess water from the blood. The other method filters the blood internally, using a special fluid that is inserted into the abdomen and removed when it has done its job. Another treatment for a failing kidney is a kidney transplant—where a healthy kidney is donated and implanted into a sick person's body. Unfortunately, donated kidneys are in short supply, and it can take years for a suitable kidney to become available for a transplant.

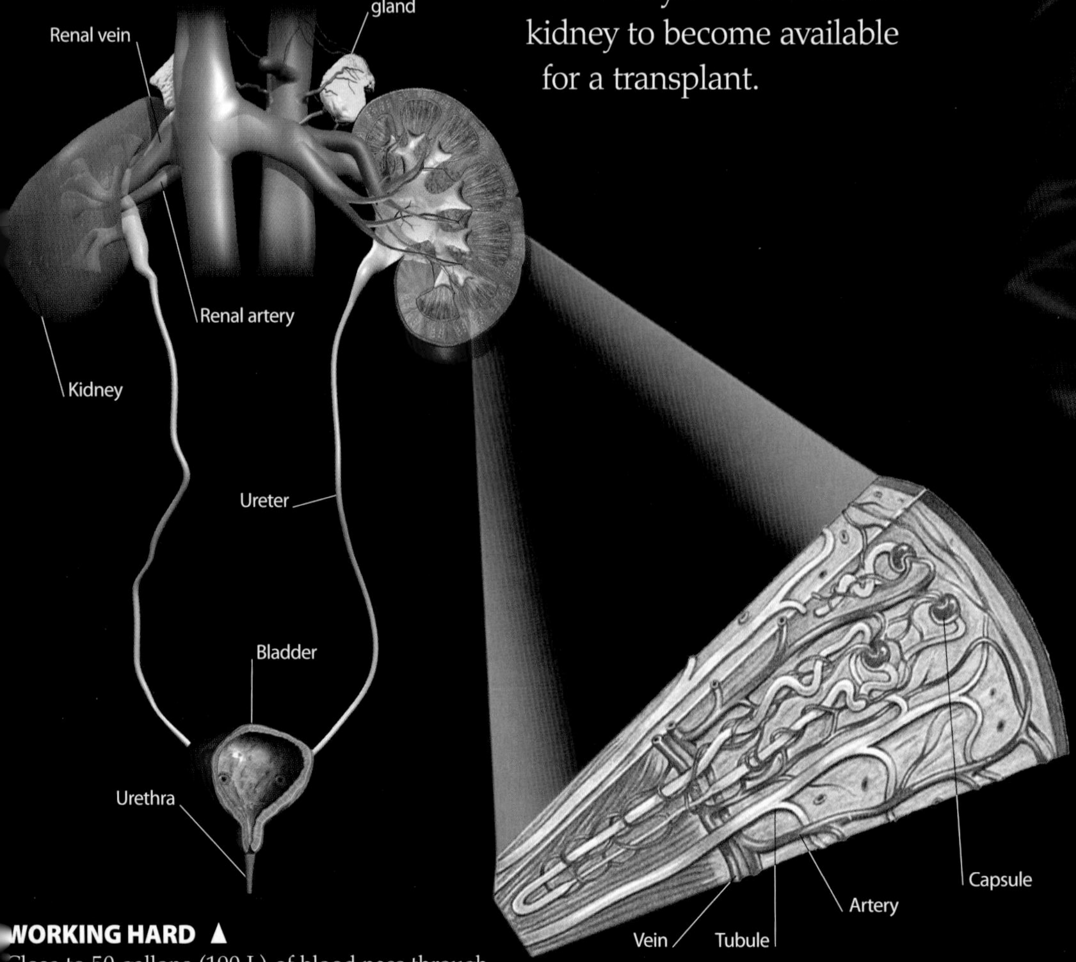

Adrenal gland

Renal vein

Renal artery

Kidney

Ureter

Bladder

Urethra

Capsule

Artery

Vein

Tubule

WORKING HARD ▲
Close to 50 gallons (190 L) of blood pass through the kidneys each day. That's a lot of filtering for these bean-shaped organs, which are only about the size of a computer mouse. Wastes that have been removed from the blood are disposed of as urine, which exits the body through the ureters, bladder, and urethra.

TINY FILTERS ▲
Each kidney is packed with around one million little filtration units called *nephrons*, like the one shown here. In the nephron, blood flows through the capsule, where the waste is removed to the tubules and eventually disposed of as urine.

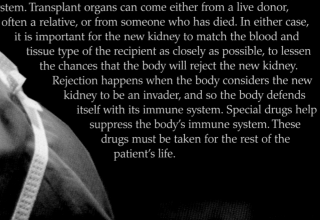

▼ CHANGING A LIFE

These surgeons are performing a kidney transplant operation. The complex process takes around three hours. Most often, the patient's own kidneys are left in place. The donated organ is inserted into the abdomen and connected into the excretory system. Transplant organs can come either from a live donor, often a relative, or from someone who has died. In either case, it is important for the new kidney to match the blood and tissue type of the recipient as closely as possible, to lessen the chances that the body will reject the new kidney. Rejection happens when the body considers the new kidney to be an invader, and so the body defends itself with its immune system. Special drugs help suppress the body's immune system. These drugs must be taken for the rest of the patient's life.

Human babies begin life inside the mother. This period of gestation is called *pregnancy*. It lasts about 40 weeks, or 9 months. For the first 8 weeks of pregnancy, the developing baby is called an *embryo*. After 8 weeks, it is called a *fetus*. The developing baby grows within an amniotic sac in the mother's uterus. This sac is filled with fluid that protects and cushions the baby. An umbilical cord connects the baby to the placenta, a thick cushion of tissue that provides a constant supply of nutrients and oxygen from the mother. The placenta allows a developing baby to stay within the safety of the mother's body for 9 months. This gives the heart, lungs, and other organs time to develop fully so they can function on their own at the time of birth.

A HUMAN FETUS AT 31 WEEKS ▶

At 4 weeks, the brain, spinal cord, and heart have begun to form. By 8 weeks, the heart is beating steadily, and at 12 weeks, the fetus can make a fist. By 20 weeks, the fetus can hear, swallow, and even scratch itself with tiny fingernails. By 31 weeks, it kicks and jabs. Its bones are soft, but fully formed, and it weighs around 3.5 pounds (1.6 kg).

did you know? ....
THE HEAVIEST NEWBORN RECORDED WEIGHED MORE THAN 22 POUNDS (10 KG). THE LIGHTEST TO SURVIVE WEIGHED 8.6 OUNCES (244 G).

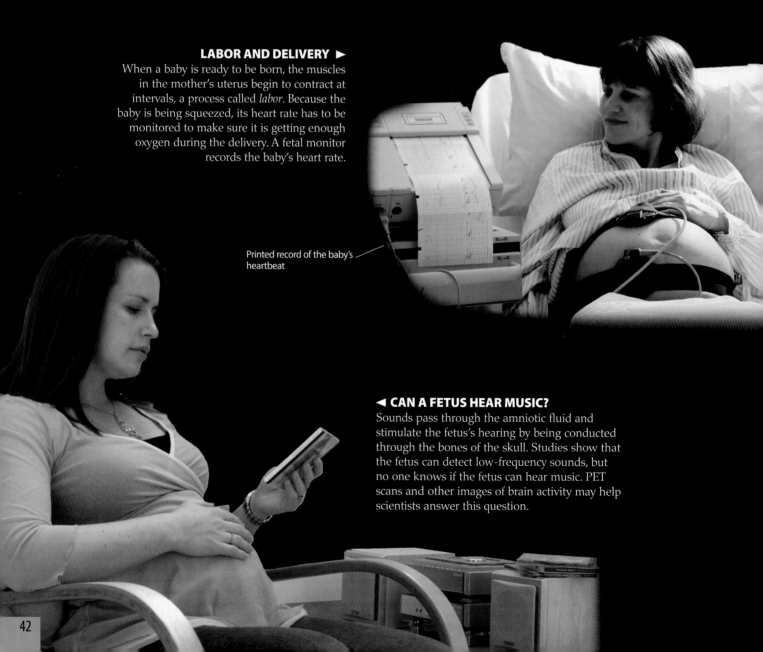

LABOR AND DELIVERY ▶

When a baby is ready to be born, the muscles in the mother's uterus begin to contract at intervals, a process called *labor*. Because the baby is being squeezed, its heart rate has to be monitored to make sure it is getting enough oxygen during the delivery. A fetal monitor records the baby's heart rate.

Printed record of the baby's heartbeat

◀ CAN A FETUS HEAR MUSIC?

Sounds pass through the amniotic fluid and stimulate the fetus's hearing by being conducted through the bones of the skull. Studies show that the fetus can detect low-frequency sounds, but no one knows if the fetus can hear music. PET scans and other images of brain activity may help scientists answer this question.

The lanugo, a fine, downy covering of hair, covers the fetus at around 20 weeks old. It begins to fall off at around 32 weeks.

By 32 weeks, the fetus can open and close its eyes and sense changes in light.

Blood vessels in the umbilical cord transport oxygen and nutrients from the placenta to the fetus and carry away waste.

NAMING

What's in a name? A lot of information! Scientists name organisms by the characteristics they share, using a system called *binomial nomenclature.* Each organism gets two names—its genus and its species, usually in Latin or Greek. Members of a genus share most characteristics. Each species can reproduce only with another member of its species. These two-part names are an organism's ID. Take this frog, *Pseudacris triseriata,* for example. The genus name, *Pseudacris,* means "false locust," probably because it makes an insectlike sound. Its species name, *triseriata,* means "three-striped." In addition to genus and species, organisms are also grouped into larger levels of classification. For example, *Pseudacris triseriata* and all of the living things shown here are organisms whose cells have a nucleus, so their first level of classification, called *domain,* is Eukarya. This frog's next level, called *kingdom,* is Animalia, followed by its phylum, which is Chordata. The next levels are class (amphibian), order (a frog), and family (a tree frog). Finally, we get down to the levels of genus and species, a false locust with three stripes on its back, commonly called a striped chorus frog.

Pseudacris triseriata

The genus name *Euglena* comes from two Greek roots that mean "good eye." *Euglena* have an eye spot that helps them sense light.

THE PROTIST AND THE FUNGI

You may have seen a member of the genus *Euglena* when you looked at a drop of pond water through a microscope. The one above is a model of one of more than 200 species in the genus. *Euglena* are in the kingdom Protista, the protists, which are organisms that are neither animal nor plant. *Euglena* are pulled through the water by their tail-like flagella, yet have plantlike structures that enable them to make their own food. Fungi, such as mushrooms, bread mold, and ring worm, have their own kingdom too, because unlike plants, they cannot make their own food.

▼ MEMBERS OF THE PLANT KINGDOM

Organisms in the Plantae kingdom produce their own food through the process of photosynthesis. They range from tiny rootless mosses to giant redwoods. The largest and most diverse phylum in the kingdom is Magnoliophyta, also known as angiosperms or flowering plants. The flowering plants are divided into monocots and dicots according to the type of seed the plant produces.

The classification of seaweeds, which are a type of multicellular algae, is in flux. Some systems place the phylum Rhodophyta, or red algae, in the plant kingdom; others place it in the protist kingdom; and some even say Rhodophyta should form its own kingdom. There are thousands of species of red algae, many of which are edible.

did you know?
A SPECIES OF DINOSAUR DISCOVERED IN 2006 WAS NAMED *DRACOREX HOGWARTSIA*, MEANING DRAGON KING OF HOGWARTS, BECAUSE IT LOOKED LIKE SOMETHING HARRY POTTER MIGHT HAVE MET UP WITH AT HIS SCHOOL FOR WIZARDS.

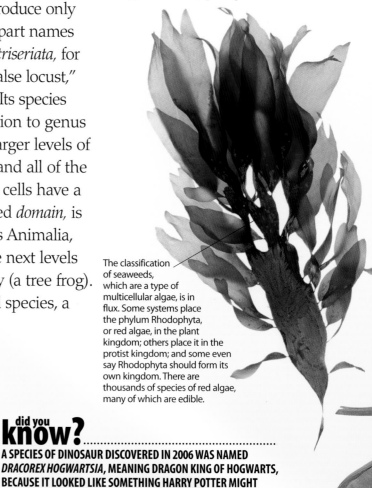

Amanita (fungus) *muscaria* (housefly) may have gotten this name because it used to be sprinkled on milk to attract flies, which would die from its toxins.

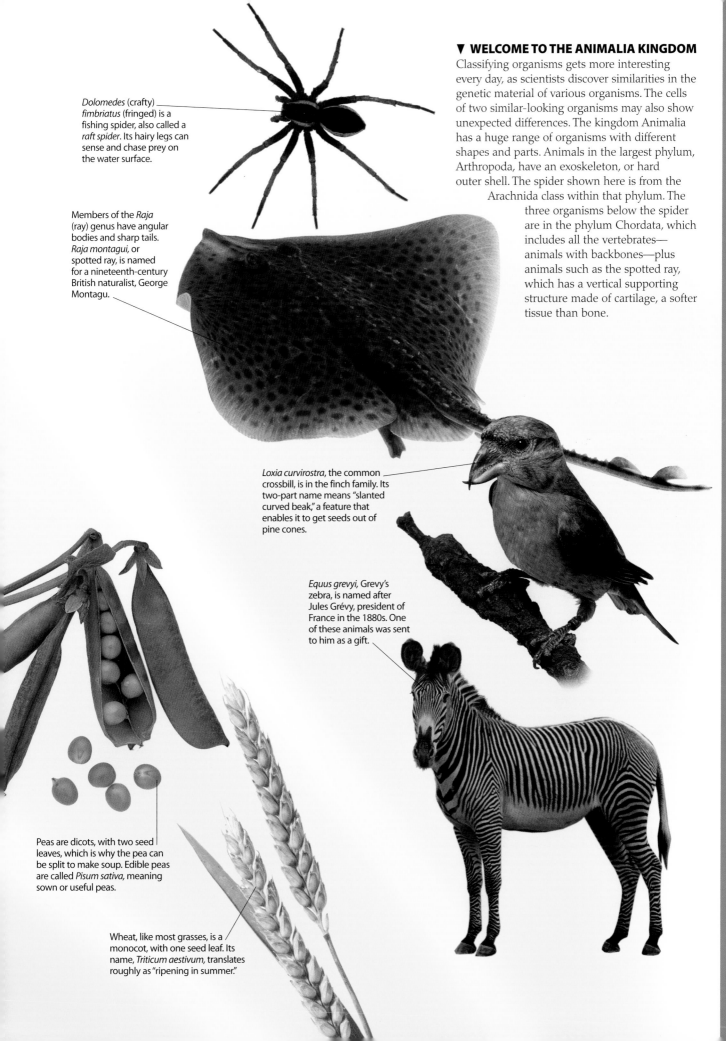

Dolomedes (crafty) *fimbriatus* (fringed) is a fishing spider, also called a *raft spider*. Its hairy legs can sense and chase prey on the water surface.

▼ WELCOME TO THE ANIMALIA KINGDOM

Classifying organisms gets more interesting every day, as scientists discover similarities in the genetic material of various organisms. The cells of two similar-looking organisms may also show unexpected differences. The kingdom Animalia has a huge range of organisms with different shapes and parts. Animals in the largest phylum, Arthropoda, have an exoskeleton, or hard outer shell. The spider shown here is from the Arachnida class within that phylum. The three organisms below the spider are in the phylum Chordata, which includes all the vertebrates— animals with backbones—plus animals such as the spotted ray, which has a vertical supporting structure made of cartilage, a softer tissue than bone.

Members of the *Raja* (ray) genus have angular bodies and sharp tails. *Raja montagui*, or spotted ray, is named for a nineteenth-century British naturalist, George Montagu.

Loxia curvirostra, the common crossbill, is in the finch family. Its two-part name means "slanted curved beak," a feature that enables it to get seeds out of pine cones.

Equus grevyi, Grevy's zebra, is named after Jules Grévy, president of France in the 1880s. One of these animals was sent to him as a gift.

Peas are dicots, with two seed leaves, which is why the pea can be split to make soup. Edible peas are called *Pisum sativa*, meaning sown or useful peas.

Wheat, like most grasses, is a monocot, with one seed leaf. Its name, *Triticum aestivum*, translates roughly as "ripening in summer."

BACTERIA

When people say bacteria are everywhere, they really mean *everywhere!* Many bacteria are what's called *extremophiles.* An extremophile is a living thing that can survive under severe conditions. The Greek root *-phil-* means love, and these organisms love extreme places. Some bacteria are acidophiles, which means they live in acids. Other bacteria are halophiles and need to live in very salty water—so salty that it would kill most other living things. Xerophiles include bacteria that can live in rocks and soils that have very little water. Some bacteria—called *psychrophiles*—can even endure freezing temperatures. They live in polar ice caps. Because of the many different ways that they can survive, bacteria live in practically every environment on Earth. Scientists study extremophiles to see how life might form on other planets, where conditions are extreme.

HOT STUFF ▶

Some like it hot—some bacteria that is. Water heated by melted rock deep underground reaches skin-burning temperatures in the Grand Prismatic Spring in Yellowstone National Park in Wyoming. Most organisms cannot survive such heat, but thermophiles—heat-loving bacteria—make this natural hot spring their home. The colorful bacteria that live there can survive in water as hot as 167°F (75°C).

❶ HEATED GROUNDWATER

The mineral water at the deepest center of the hot spring is 188°F (87°C). No life can survive here.

❷ COOLING OFF

Water in the shallower parts of the hot spring has been cooled slightly by the surrounding air. Bacteria that are green in color can survive the cooler temperatures.

❸ COOL AND COLORFUL

The bacteria that form the green, brown, and yellow slime are called *cyanobacteria.* They are yellow-green in warmer temperatures.

❹ NOT SO HOT

As waters cool near the edge of the hot spring, the cyanobacteria become orange and yellow. The color pigments produced by the bacteria act as a sunscreen.

❺ DRYING OUT

The soil surrounding the hot spring is dry and cool—a less comfortable home for thermophiles.

VIRUS ATTACK ►

All living things can catch a virus, even bacteria. When viruses attack bacteria cells, they don't actually enter the cell. Instead, they inject only their DNA. The cell treats the viral DNA as if it were its own, using it to create viral parts instead of bacteria parts and forcing it to become a tiny factory for making more viruses.

Viruses that attack bacteria are called *bacteriophages.*

Virus attacks often result in the death of the cell.

A cell that is attacked by a virus is called a *host.*

◄ BACTERIA CELL

Bacteria are the smallest living things on Earth. They live as single cells, so each cell must carry out all the functions of life.

Bacteria whip their thread-like flagella to move through their liquid surroundings.

The tough cell wall offers protection.

The flexible cell membrane allows nutrients in and wastes out.

The cell's cytoplasm holds its DNA and other molecules important for life.

did you know?
THE NUMBER OF BACTERIA ON YOUR BODY RIGHT NOW IS GREATER THAN THE NUMBER OF PEOPLE IN THE UNITED STATES.

MOLD

Have you ever opened the refrigerator for a snack only to find nasty green fur, gray splotches, orange fuzz, or white dusty stuff on something you were hoping to eat? This is mold, a type of fungus. Mold doesn't just grow on food; it can also grow on plants, paper, and even walls. Like the vast body of an iceberg sitting below the surface of the water, most of a mold is hidden under the surface on which it grows. The part of the mold that you can see contains microscopic spores that will be released into the air by the millions. Mold spores are all around you—in the air that you breathe, and on virtually every surface. Some molds can cause allergic reactions and respiratory infections. They thrive in damp, warm areas, but spores can survive almost anywhere until the conditions are just right for them to grow!

BATTLE OF THE MOLD! ▼

The thousands of homes and businesses flooded by Hurricanes Katrina and Rita in 2005 provided the perfect conditions for mold to grow. Because buildings stayed wet for so long, mold took over, covering any surfaces it could. Efforts to remove the mold were not always successful.

did you know?
A STUFFY NOSE AND ITCHY EYES MIGHT BE AN ALLERGIC REACTION TO MOLD. MOLDS CAN ALSO CAUSE ASTHMA ATTACKS.

Mold spores in high concentrations can cause health problems. Face masks and gloves were necessary to protect workers during the massive cleanup operation.

Carpets, cupboards, tables, walls, and floors—all were attacked by molds after the flooding.

NOT QUITE SO PEACHY! ▶

It's bad enough when mold invades your fruit bowl, but on a large scale it can be a farmer's worst enemy. Crops can be wiped out by mold if storage conditions are not right or if the weather turns damp too early. However, some molds are useful in food production. Blue-veined cheeses like Stilton get their coloring and taste from mold. Mold is also used when making soy sauce and tempeh.

Stilton cheese

A mold changes in both color and texture as the spores ripen.

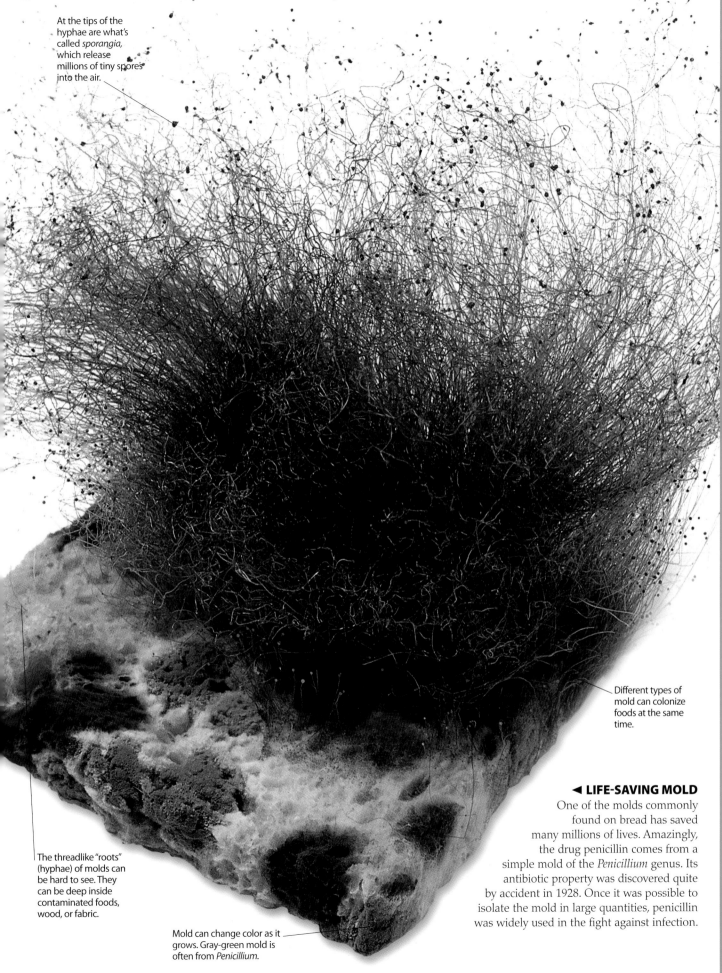

At the tips of the hyphae are what's called *sporangia*, which release millions of tiny spores into the air.

Different types of mold can colonize foods at the same time.

The threadlike "roots" (hyphae) of molds can be hard to see. They can be deep inside contaminated foods, wood, or fabric.

Mold can change color as it grows. Gray-green mold is often from *Penicillium*.

◄ LIFE-SAVING MOLD

One of the molds commonly found on bread has saved many millions of lives. Amazingly, the drug penicillin comes from a simple mold of the *Penicillium* genus. Its antibiotic property was discovered quite by accident in 1928. Once it was possible to isolate the mold in large quantities, penicillin was widely used in the fight against infection.

FUNGI

Fungi are such unique organisms that they form their own scientific kingdom, whose members range from unicellular organisms such as yeast to multicellular organisms that can look similar to plants. But unlike plants, fungi don't have the chlorophyll needed to absorb sunlight and make nutrients. Instead, fungi secrete enzymes onto the surfaces where they grow, such as on wood, plants, fruit, or even dung. The enzymes cause the surface to decompose, and the fungi absorb the nutrients through a system of tiny threadlike cells, called the *mycelium*. When we see fungi, like the head and stem of a mushroom, or mold growing on bread, what we see is usually what's called the *fruiting body*. Often more than 90 percent of the fungus is composed of the underground mycelium. Fungi can live almost anywhere, including on plants and animals. Many plants have a symbiotic relationship with fungi. The plant gives the fungus some energy through photosynthesis. In return, the fungus helps the plant take up water and minerals.

TURKEY TAILS ▶

These leathery bracket fungi—shelflike growths—do resemble their name. They grow by breaking down dead wood. The result of this process is that nutrients from the tree return to the soil.

FLY AGARIC—BEWARE! ▼

Fly agaric is definitely not for eating! This fungus has been used to kill flies by attracting them to milk that has small pieces of the poisonous fungus soaking in it. The individual threadlike cells of fly agaric's mycelium, called *hyphae*, tap into tree roots to obtain nutrients.

Gills, where millions of reproductive structures called *spores* are produced

The scarlet wax cap mushroom is brightly colored and feels waxy.

The mushroom's stem is called a *stape*.

POWDER-FILLED PUFFBALLS ▶

The fruiting body of a puffball can be as tiny as a large pea or bigger than a watermelon. Calculations have shown that a soccer-ball sized giant puffball contains about 7 trillion spores. The largest giant puffball on record weighed in at a mighty 48 pounds (almost 22 kg) and was more than 8 feet (more than 2.6 m) in diameter!

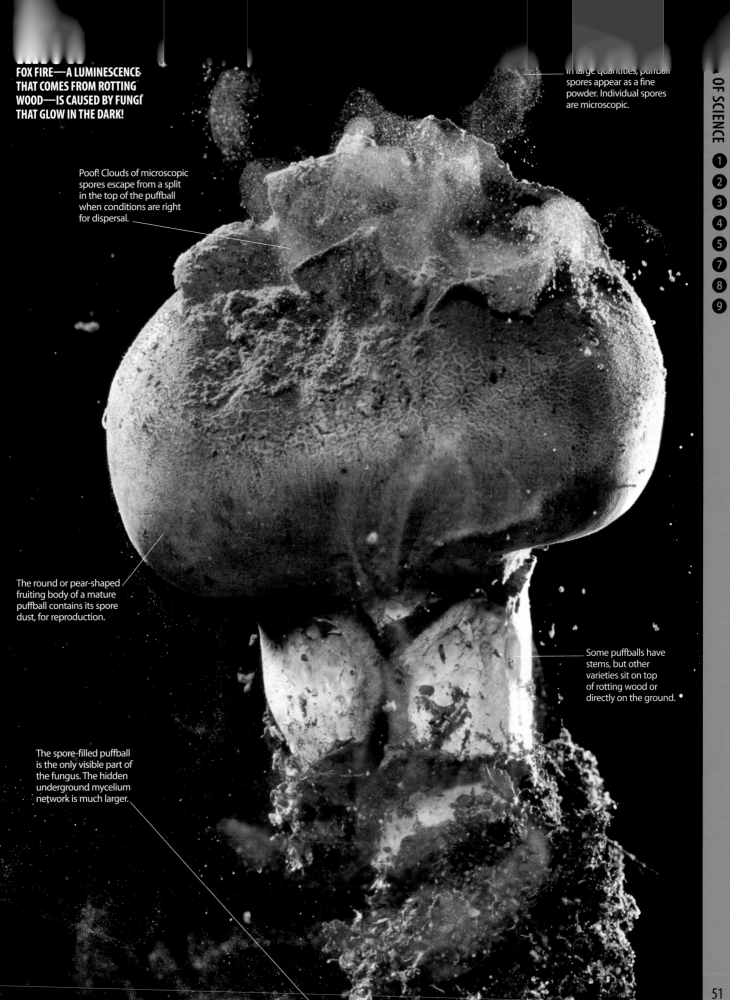

FOX FIRE—A LUMINESCENCE THAT COMES FROM ROTTING WOOD—IS CAUSED BY FUNGI THAT GLOW IN THE DARK!

In large quantities, puffball spores appear as a fine powder. Individual spores are microscopic.

Poof! Clouds of microscopic spores escape from a split in the top of the puffball when conditions are right for dispersal.

The round or pear-shaped fruiting body of a mature puffball contains its spore dust, for reproduction.

Some puffballs have stems, but other varieties sit on top of rotting wood or directly on the ground.

The spore-filled puffball is the only visible part of the fungus. The hidden underground mycelium network is much larger.

ALLERGIES

After Jamie took a sip of Liza's drink, his throat swelled and he was rushed to the hospital. Six-month-old Kate was there because she had started wheezing. What was the culprit? Allergies. Liza had been eating peanuts, and Jamie, who is allergic to nuts, had experienced a severe allergic reaction to Liza's saliva on the cup. Kate's asthma was caused by her allergies to the pollens in the air. An allergy is a reaction by your immune system to something that does not bother most people. Allergens, the substances that cause allergies, are everywhere—pollens, foods, mold, dust mites, pets, even medicines—and about one in five people in the United States suffers from allergies. Some allergic reactions can be irritating, such as sneezing, hives, or watery eyes, but some can be deadly if they are not treated immediately.

DUST MITES ▶

Unlike this model, real dust mites can't be seen without magnification. Yet they can cause big allergy problems—sneezing, itching, watery eyes, stuffy ears, skin rashes, and even asthma. Dust mite feces and tiny pieces of their bodies get mixed in with dust. When the dust is disturbed, the particles can be inhaled. Achoo!

Dust mites are eight-legged creatures that are related to spiders.

Some mites' hairs may sense electrical signals.

Dust mites feed on dead skin cells that people shed.

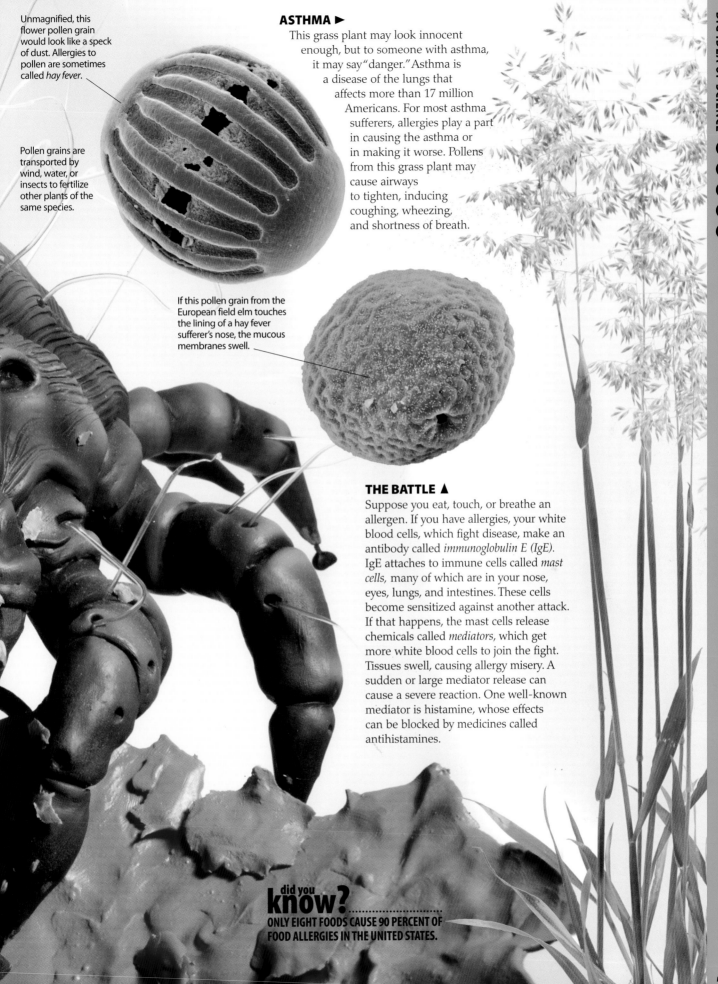

Unmagnified, this flower pollen grain would look like a speck of dust. Allergies to pollen are sometimes called *hay fever*.

Pollen grains are transported by wind, water, or insects to fertilize other plants of the same species.

ASTHMA ▶

This grass plant may look innocent enough, but to someone with asthma, it may say "danger." Asthma is a disease of the lungs that affects more than 17 million Americans. For most asthma sufferers, allergies play a part in causing the asthma or in making it worse. Pollens from this grass plant may cause airways to tighten, inducing coughing, wheezing, and shortness of breath.

If this pollen grain from the European field elm touches the lining of a hay fever sufferer's nose, the mucous membranes swell.

THE BATTLE ▲

Suppose you eat, touch, or breathe an allergen. If you have allergies, your white blood cells, which fight disease, make an antibody called *immunoglobulin E (IgE)*. IgE attaches to immune cells called *mast cells*, many of which are in your nose, eyes, lungs, and intestines. These cells become sensitized against another attack. If that happens, the mast cells release chemicals called *mediators*, which get more white blood cells to join the fight. Tissues swell, causing allergy misery. A sudden or large mediator release can cause a severe reaction. One well-known mediator is histamine, whose effects can be blocked by medicines called antihistamines.

did you know?..........................
ONLY EIGHT FOODS CAUSE 90 PERCENT OF FOOD ALLERGIES IN THE UNITED STATES.

PANDEMIC

Pathogens—bacteria, viruses, fungi, and protists—cause infectious diseases. These diseases are spread in different ways: through water, food, or soil; via animals; or by human contact. An epidemic occurs when an infectious disease spreads quickly to many people throughout a country or region. An epidemic that spreads rapidly to even more people in different parts of the world is called a *pandemic*. Some diseases that have historically become pandemics are influenza (flu), cholera, and plague. These and other infectious diseases can spread throughout the world more quickly than ever before, due to great increases in international travel. The World Health Organization (WHO) keeps track of infectious diseases around the world and decides when they reach epidemic and pandemic levels. WHO has devised a pandemic preparedness plan that outlines steps that should be taken in the event of an influenza pandemic.

◀ CHOLERA

The pathogen that causes cholera in humans is a type of bacteria. The bacteria enter the intestine when a person drinks contaminated water or eats contaminated food. In severe cases, cholera causes vomiting and diarrhea and, left untreated, can lead to death. In the United States, cholera is not a concern because of the advanced water treatment systems.

Cholera is often associated with watery diarrhea. The bacteria actually produce a cholera toxin that causes this symptom.

◀ IS THE WATER CLEAN?

Some pathogens, such as the bacteria that cause cholera, can live in water. When a healthy person drinks contaminated water, he or she ingests these pathogens and can become ill. Microorganisms can be introduced into groundwater in various ways. For example, fecal contamination from sick animals can seep into groundwater. A leak in a wastewater system can also contaminate water with pathogens. Areas of the world with poor water treatment systems often spur the growth of pandemics because the pathogens may infect great numbers of people in a short time.

In Afghanistan, conflict and drought have affected the cleanliness of the water supply.

NOVEL H1N1 VIRUS ▶

Novel, meaning new, is the best description for this virus that was first seen in humans in the spring of 2009. On June 11, 2009, the World Health Organization officially declared it a pandemic. H1N1 spreads quickly between humans in the same way other flu viruses do—through coughing and sneezing. As opposed to typical flu viruses, H1N1 is harder on people younger than 25 years old than it is on the elderly. Vaccines against H1N1 have been developed.

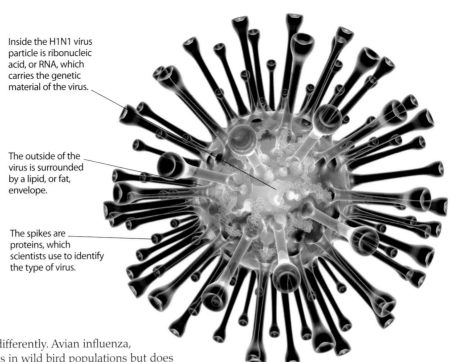

Inside the H1N1 virus particle is ribonucleic acid, or RNA, which carries the genetic material of the virus.

The outside of the virus is surrounded by a lipid, or fat, envelope.

The spikes are proteins, which scientists use to identify the type of virus.

THE BIRD FLU ▼

Viruses affect different animal populations differently. Avian influenza, commonly known as bird flu, naturally exists in wild bird populations but does not usually make them sick. But bird flu is very contagious between birds. It can be passed to domesticated birds such as chickens and turkeys through saliva, nasal secretions, and feces. In these birds, the virus has caused millions of deaths. Bird flu has crossed over to humans mostly in cases where people have had contact with bodily fluids of infected birds. As with all new viruses, scientists work to develop vaccines. Bird flu vaccines are now being given to domestic bird populations to try to stop the spread of the disease.

did you know?..........................
THERE HAVE BEEN AT LEAST SEVEN CHOLERA PANDEMICS SINCE THE EARLY NINETEENTH CENTURY.

Suppose a virus enters your body. This disease-causing pathogen has certain molecules, called *antigens,* all around its outside surface. White blood cells called *lymphocytes* recognize the virus by its antigens. Soon, lymphocytes called B cells begin producing Y-shaped particles, called *antibodies.* Antibodies are like puzzle pieces designed to lock onto specific antigens. Once the antibodies lock on, the viruses can't attack your body's cells. If the same type of virus enters your body again, your immune system "remembers" how to defeat it. This process gives your body what's called *active immunity* against that pathogen. Vaccines cause your body to develop active immunity to a disease without causing you to have the disease. A vaccine includes virus antigens that have been killed or weakened. They cannot cause the disease. The immune system, however, produces antibodies just as it would if the antigens were dangerous. As a result, a vaccinated person exposed to the disease is very unlikely to get sick.

TESTING A VACCINE ▼

Before a vaccine can be used widely, it must be tested for safety and effectiveness. This woman is receiving a shot of a vaccine against the H1N1 virus, sometimes called "swine flu." About 8 to 10 days later, researchers tested her blood to see if antibodies against the H1N1 virus had developed. Antibodies had developed, showing the vaccine to be effective.

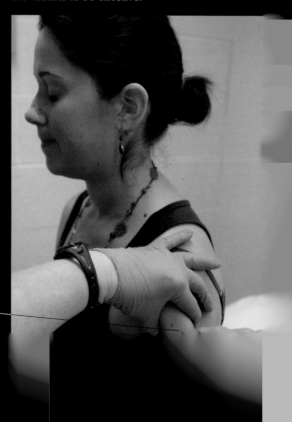

MASS IMMUNIZATION ▲

An epidemic—a widespread, sudden outbreak of a disease—often leads a country's government to set up a mass immunization program, in which everyone living in an area receives a vaccine. The vaccinations are usually given first to the people most likely to get sick and to those most likely to be exposed to the disease, such as health care workers. Vaccination programs have eliminated diseases, such

Influenza, or flu, shots need to be given yearly. Flu vaccines typically are changed from year to year to include the viruses that scientists predict will be circulating each year.

Both smallpox and cowpox viruses are members of the *Orthopoxvirus* genus. Vaccines are made using another member of the genus.

did you know?............................

VACCINE COMES FROM A GREEK WORD FOR COW, BECAUSE OF JENNER'S WORK WITH COWPOX.

The smallpox virus's protein coat, shown in green, protects its genetic material and enables the virus to penetrate a person's body cells.

Viruses infect body cells and use the cell to reproduce their genetic material (red) and to produce more viruses.

The smallpox virus exists now only in a few research laboratories.

▲ SMALLPOX VIRUS

English physician Edward Jenner noticed that people who had suffered from the mild disease cowpox didn't contract the deadly disease smallpox. In 1796, Jenner rubbed infectious material from a woman's cowpox sores into scratches on a boy's arm. The boy became ill with cowpox. Later, Jenner exposed him to smallpox, and the boy did not get sick. He had developed immunity to this group of related viruses. Jenner's discovery has since led to the wiping out of smallpox, a killer of millions of people.

COMMON COLD

"Ah…ah….ah…Achoo! Oh, do. I tink I hab a code!" You know it when you feel it: a headache, sore throat, stuffed up nose, sneezing, coughing, and the seemingly endless river of snot. You have a cold! More than 200 different viruses can cause the common cold. Because so many different germs are responsible for this familiar illness, scientists and doctors have little hope of finding an effective cure anytime soon. The easiest way to take care of a cold is to never catch it in the first place. Keep clear of uncovered sneezes, and wash your hands regularly. Soap and water will kill a cold virus, which can otherwise survive for hours on surfaces such as doorknobs, railings, drinking cups, money, and skin.

DAY 3: SORE THROAT AND MILD FEVER

The sore throat that you feel is not from the death of cells destroyed by the viruses. Instead, the discomfort comes from the immune system signals produced by your own body. They cause swelling of the tissues and trigger pain-sensing nerve cells. Other immune system signals cause fever and aching pain in your head and muscles.

THE STAGES OF A COLD ▶

A cold can last one to two weeks and may make you feel downright miserable. But while a virus is to blame for the infection, it is your body's immune response that causes the symptoms you have come to dread.

DAYS 1 & 2: INFECTION

You catch a cold when viruses enter your nose or throat and start to multiply using your own cells as tiny viral factories. But, who gave you those viruses? It's not always easy to tell. It takes two or three days for any cold symptoms to appear. You can spread viruses to others within a day of infection and for as many as three days after you no longer feel sick.

did you know? A SNEEZE CAN HAVE THE WIND SPEED OF A CATEGORY 2 HURRICANE.

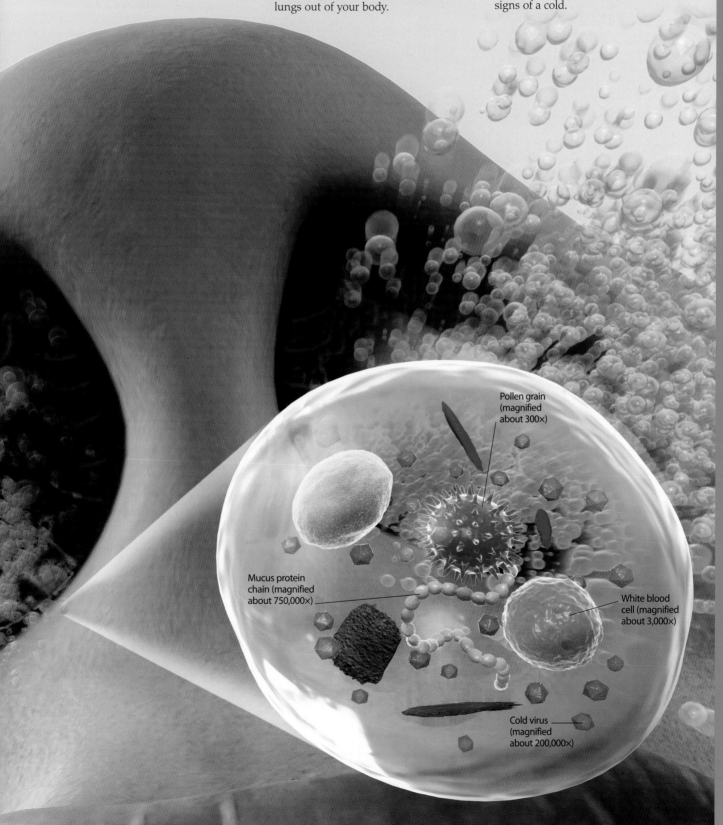

DAYS 4 & 5: RUNNY NOSE AND SNEEZING

Mucous glands in your nose start making mucus, or snot, a slippery liquid that contains water and proteins. Snot may start out clear and colorless but soon becomes green due to the presence of a green coloring in white blood cells that are fighting your infection. Snot helps wash the viruses out of your body, and sneezing helps blast them out at a faster pace.

DAYS 6 & 7: STUFFINESS AND COUGHING

The stuffed up feeling that makes it hard to breathe is not from all the snot. It is due to the dilation of blood vessels in the nose tissues, also caused by immune system signals. When the infection gets far down into your throat, the irritated nerves cause coughing. Coughing can help move infected mucus that has built up in the lungs out of your body.

DAY 20: PROBABLY NOT A COLD

If you are sneezing for more than two weeks, see a doctor. A cold might not be your problem. An allergen like pollen is more likely to blame. Some sneezers mistake allergies for colds. Many of the symptoms are the same, but colds do not usually cause itchiness in the eyes and nose as allergies often do. Also, allergy symptoms do not usually include the achy fever and colorful snot that are signs of a cold.

Pollen grain (magnified about 300×)

Mucus protein chain (magnified about 750,000×)

White blood cell (magnified about 3,000×)

Cold virus (magnified about 200,000×)

MALARIA

It may feel like the flu—a high fever, head and muscle aches, tiredness, and chills. But if left untreated, malaria can be a deadly disease. Malaria is caused by a tiny parasite. The parasite infects a particular kind of mosquito. The mosquito carrying the parasite bites humans, transmitting the disease. One way to prevent malaria is by protecting against mosquito bites in areas where the disease occurs. A person who becomes sick with malaria can be treated with prescription medication. The earlier the treatment begins, the more likely the person will recover. With the right medication, people who have malaria can be cured. But, the best way to combat malaria is to prevent it.

Mosquito eyes have excellent night vision.

Blood is sucked through the proboscis.

◄ SPREADING MALARIA

You cannot get malaria by being near or touching someone who has it. You usually have to be bitten by a mosquito, and not just any mosquito—a female *Anopheles* mosquito. When a mosquito bites a person who has malaria, the mosquito takes in that person's blood and becomes infected. When that same mosquito bites a second person, the mosquito injects a mix of its infected blood and its saliva into that person.

did you know?
BETWEEN 350 AND 500 MILLION CASES OF MALARIA OCCUR IN THE WORLD EACH YEAR, CAUSING OVER ONE MILLION DEATHS.

MALARIA AROUND THE WORLD ►
Malaria is most common in tropical and subtropical areas of the world. Scientists have been working on a malaria vaccine for more than 50 years. The first Phase 3 clinical trial of a malaria vaccine began in May 2009 in seven African countries.

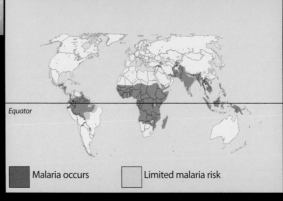

Equator

Malaria occurs   Limited malaria risk

PREVENTING MALARIA ▲

Prevention is the key to stopping
the spread of malaria. One of the
most important tools in this fight is the
mosquito net. People living in or visiting areas
where malaria exists should sleep under mosquito
nets, preferably ones that have been treated with an
insecticide. They should also use insect repellent and wear
long-sleeved clothing when outside at night.

CANCER TREATMENT

You may know of someone who died from cancer, but you may also know of someone whose cancer was cured. Cancer is a group of diseases in which abnormal cells grow and divide in an uncontrolled way. These cells can invade surrounding tissue or spread to other parts of the body. The form of cancer treatment depends on the type of cancer and whether it has spread. Some abnormal growths, or tumors, can be completely removed by surgery. If the cancer has spread, however, doctors may treat it with drugs, a process called *chemotherapy* or *chemo*. They may also treat the cancer with radiation, called *radiation therapy* or *radiotherapy*. Sometimes doctors use both treatments together to destroy cancer cells.

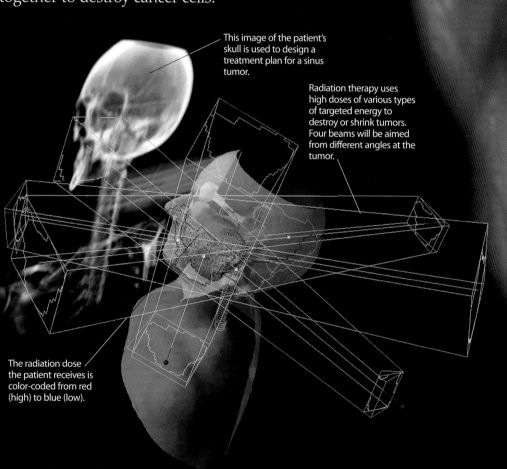

This image of the patient's skull is used to design a treatment plan for a sinus tumor.

Radiation therapy uses high doses of various types of targeted energy to destroy or shrink tumors. Four beams will be aimed from different angles at the tumor.

The radiation dose the patient receives is color-coded from red (high) to blue (low).

DESTROYING CANCER CELLS ▲

During the radiation therapy shown here, a special machine moves around the patient's body, aiming high-energy rays at the tumor. Although this treatment targets the cancer cells, some nearby healthy cells may also be damaged. In contrast, chemotherapy drugs are swallowed or injected, so the patient's whole body is exposed to them. That's why chemo usually has more side effects, such as hair loss and nausea.

did you know?................
IN THE UNITED STATES, LUNG CANCER IS THE LEADING CAUSE OF CANCER-RELATED DEATHS FOR BOTH MEN AND WOMEN.

▼ ROBODOC

Scientists are developing many promising new cancer treatments. Some researchers are creating drugs that block the formation of the blood vessels that nourish cancer cells. Other researchers are developing microscopic devices, called *nanorobots*, that are injected into a patient's body to find, diagnose, and treat cancer cells.

Powered with chemical "engines," nanorobots will not only locate cancer cells but also show whether a cancer has spread.

Scientists intend to use nanorobots to inject chemotherapy drugs directly into cancer cells, without damaging healthy cells. They could also be used to deliver radiation therapy and even to perform surgery.

Cancer cell

NUCLEAR MEDICINE

Radioactivity may signal "Danger!" in your mind. However, a branch of medicine, called *nuclear medicine*, uses radioactivity to help diagnose and treat diseases. Nuclear medicine takes advantage of the fact that radioactive substances are unstable. This means that the isotopes of certain chemical elements emit particles with high-energy values, such as gamma rays. Isotopes are atoms of the same element with different numbers of neutrons. When radioactive drugs are injected into the human body, their particles give off energy that is detected to create images of the areas where they were absorbed. This helps locate problems in the body. The drugs used in this imaging process do not harm the body. However, the gamma rays these drugs emit can be used to destroy cells. For example, harmful cancer cells are often targeted with gamma radiation. Another example is the use of radioactive drugs to treat the thyroid gland—an important gland in your body that helps regulate your metabolism.

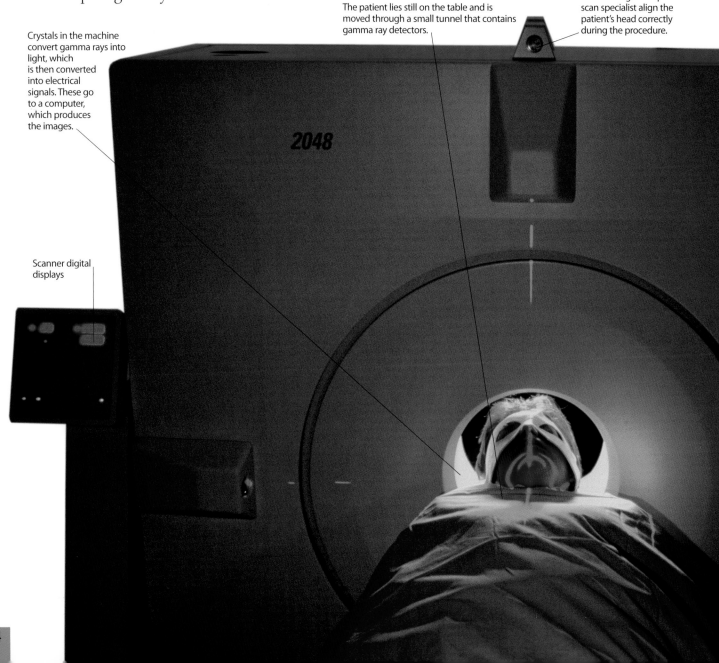

The patient lies still on the table and is moved through a small tunnel that contains gamma ray detectors.

These red lights help the scan specialist align the patient's head correctly during the procedure.

Crystals in the machine convert gamma rays into light, which is then converted into electrical signals. These go to a computer, which produces the images.

2048

Scanner digital displays

RADIO PHARMACOLOGY ▶

This chemist is blending radioactive chemicals to make radiopharmaceuticals. These are drugs that emit radioactivity. They are also called radiotracers, because their emissions can be traced. The glassed-in safety chamber shown here, along with the chemist's special clothing, will protect him from radioactivity as he mixes the chemicals.

Different chemicals are absorbed by different parts of the body. Radioactive nitrogen, for example, is used to trace blood circulation in the heart and lungs.

did you know?
UP TO 12 MILLION NUCLEAR MEDICINE IMAGING AND OTHER PROCEDURES ARE CARRIED OUT EVERY YEAR IN THE UNITED STATES.

◀ PATIENT IN PET SCANNER

This PET doesn't live in a house! PET stands for Positron Emission Tomography. Positrons are positively charged particles that are emitted from some radioactive substances. In a PET scan, a patient receives, usually through an injection, a radioactive drug that is absorbed by certain tissues of the body. As the drug passes through the body, the machine measures the gamma rays that are produced when positrons collide with electrons in the body. The patient is moved through the machine, and a series of images of the body are formed. Doctors analyze the images to look for signs of disease, such as tumors. PET scans are also used to see how organs and tissues, such as your heart and lungs, are functioning.

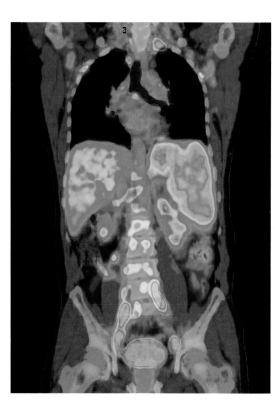

PET SCAN ▲

Patients undergoing a PET scan swallow, inhale, or are injected with a radiotracer material. The scanner detects the radioactive emissions from the injected material that enhances the image for doctors. This PET scan image shows abnormal growths in a patient's abdomen and chest caused by non-Hodgkin's lymphoma, a type of cancer that affects white blood cells. In this case, the PET scan images show doctors how far the cancer cells have developed.

BIODIVERSITY

What do you get when you combine the words *biological* and *diversity?*
Biodiversity! The word was first coined in 1985, and since then, biodiversity
has become a hot topic. Biodiversity means the variety of forms of life.
Think about how many kinds of living things there are on Earth. They
are all connected—through food chains, carbon cycles, and ties
we haven't discovered yet. Each ecosystem, all living and
nonliving things in an area, depends on its members
to maintain balance. When a species becomes
extinct, that balance is lost. Biodiversity refers
to three kinds of diversity in living things.
First, there is diversity in the types of
ecosystems around the world, such
as coral reefs and savannas.
Second, there is a variety of
species in an area, such as the
coral and plants living with
the glassfish on this reef.
Third, there is variation
within species. The
glassfish shown here all
look like bubbles, but
some individuals are
faster or longer than
others. Protecting
all three types of
biodiversity is the
key to keeping
Earth alive.

PICK AN ECOSYSTEM ▶

Hot, cold, wet, dry—you name it! There are all kinds of places where plants and animals live. Each place has a range of temperatures and a range of precipitation that is just the right combination to support the living things in that ecosystem. Take a look at these four ecosystems in West Africa.

Desert: Very few plants and animals can survive the hot, dry Sahara Desert.

Arid Grasslands: The land that borders the desert—called *sahel*, which means "shore"—is home to a few trees, low growing grasses, camels, oxen, and cattle.

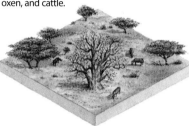

Savanna: A bit more rain here means the savanna ecosystem has more trees and grasses along with the animals, such as giraffes, that eat them.

Tropical Forest: Thousands of plants and animals live in the moist tropical forest. It is one of Earth's most complex ecosystems.

BIODIVERSITY WITHIN SPECIES ▲

This lioness stalking through the savanna is different in size, strength, and temperament, from other lions of her species. Within each species, there is diversity. This variation in traits is important because passing on helpful traits to offspring will help each species prosper.

did you know?
ABOUT 1.75 MILLION SPECIES OF LIVING THINGS ON EARTH HAVE BEEN DESCRIBED AND NAMED, AND MILLIONS MORE ARE WAITING.

◀ BIODIVERSITY BETWEEN SPECIES

Coral reefs teem with millions of species of sponges, coral, algae, and fish. It is the most diverse kind of marine ecosystem, and each species in the coral reef ecosystem interacts with the others. The health of an ecosystem is often measured through its biodiversity. The existence of many species in an ecosystem, with good variation within each species, probably means that each species has plenty of food and shelter.

The sea anemone provides protection for the clownfish. The clownfish eats parasites off the sea anemone and helps circulate water around the anemone.

This clownfish can live among the stinging tentacles of the sea anemone. It is covered with a slimy mucus that protects it from the anemone's poison.

ADAPTATIONS

Fish that puff up like balloons, spiders that pose as ants, plants that eat meat: sometimes the adaptations that allow an organism to survive can be quite bizarre. But adaptations help plants and animals obtain food and water, keep safe, establish territory, withstand weather, and reproduce. We do not usually think of plants as meat-eaters, but some feed on insects, invertebrates, and even small mammals. These plants live in poor soils with few nutrients. They have special adaptations to attract, trap, kill, and digest their prey. These include sticky surfaces, and hinged leaves that snap shut when trigger hairs are touched. Anglerfish live in the deep ocean where it is totally dark. The females have a special spine on their dorsal fin, which hangs over their mouth and acts as a "fishing rod." The fleshy tip of this lure is often luminous to attract other fishes. Munch!

There is enough toxin in one of these fish to kill 30 adults—and there is no known antidote!

PUFFED AND POISONOUS ▶
Puffer fish and porcupine fish are slow swimmers, so they've had to develop other ways to stay safe from fast swimming predators. How? These impressive fish inflate themselves when they're threatened. They have incredibly elastic skin and can balloon up to about three times their normal size. Try swallowing this fish now, predator!

Porcupine fish are also armed with sharp spines from head to tail, making them even more difficult to attack.

Distinctive white spots help identify and camouflage the highly poisonous death puffer on the reef.

Special muscles rapidly pump water into the stomach to inflate the fish. The stomach can swell to 100 times its normal size.

Adaptations develop over generations, not during an individual's life.

The Venus' flytrap stays closed for about 10 days while it digests an insect.

Huge 18-inch (46-cm) tongues and extra thick saliva allow giraffes to eat their favorite leaves from thorny acacia trees.

Giraffes have special one-way valves in their long necks to regulate blood flow and stop them from fainting.

REACHING FOR THE SKY ▲

Giraffes are the tallest animals of all, with long legs and necks that allow them to reach leaves up to 20 feet (6 m) high. To keep them from tipping over, giraffes have an unusual gait. They move the two legs on one side of the body forward, and then follow with the two legs on the other side. The giraffe's unusual shape also makes it awkward for them to lower their heads to drink, so they are more vulnerable to predators while they're drinking. Therefore, giraffes drink only about once a day, up to 10 gallons (40 L) each time.

did you know? SOME DESERT GECKOS HAVE FRINGED TOES THAT KEEP THEM FROM SINKING INTO SAND.

PATTERNS IN NATURE

Patterns are one of nature's adaptations for survival. Take stripes, for example. A tiger's stripes camouflage it in tall grass. A skunk's stripes may warn off predators. A harmless snake's red, black, and yellow stripes may mimic those of the poisonous coral snake. Symmetry is another type of pattern. Sea stars and jellyfish have radial symmetry, as a pie does. This shape lets organisms whose food floats around them sense it from any direction. Organisms that need to move around to find their food are typically bilaterally symmetrical—two halves that are mirror images—and have most of their sensory organs in their head. They can sense danger with their heads and coordinate balanced and fast movement. Certain patterns in nature can even be described mathematically by a sequence of numbers called *Fibonacci numbers*. The number of petals on some flowers, the number of spirals on a pine cone, and some say even the proportions of the human body can be found in this pattern of numbers.

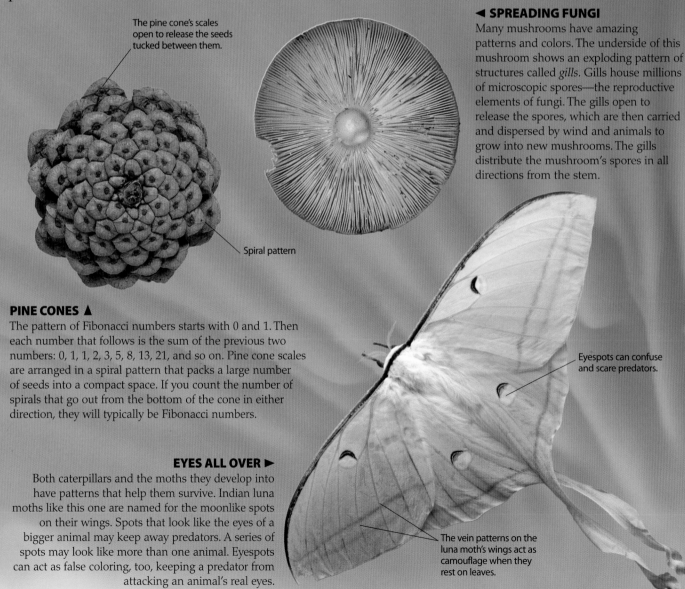

The pine cone's scales open to release the seeds tucked between them.

Spiral pattern

◀ SPREADING FUNGI
Many mushrooms have amazing patterns and colors. The underside of this mushroom shows an exploding pattern of structures called *gills*. Gills house millions of microscopic spores—the reproductive elements of fungi. The gills open to release the spores, which are then carried and dispersed by wind and animals to grow into new mushrooms. The gills distribute the mushroom's spores in all directions from the stem.

Eyespots can confuse and scare predators.

PINE CONES ▲
The pattern of Fibonacci numbers starts with 0 and 1. Then each number that follows is the sum of the previous two numbers: 0, 1, 1, 2, 3, 5, 8, 13, 21, and so on. Pine cone scales are arranged in a spiral pattern that packs a large number of seeds into a compact space. If you count the number of spirals that go out from the bottom of the cone in either direction, they will typically be Fibonacci numbers.

EYES ALL OVER ▶
Both caterpillars and the moths they develop into have patterns that help them survive. Indian luna moths like this one are named for the moonlike spots on their wings. Spots that look like the eyes of a bigger animal may keep away predators. A series of spots may look like more than one animal. Eyespots can act as false coloring, too, keeping a predator from attacking an animal's real eyes.

The vein patterns on the luna moth's wings act as camouflage when they rest on leaves.

◀ FIERCE FACE

The puss moth is a soft, fuzzy creature like the cat it is named after. But the caterpillar that the moth grows from is not! The caterpillar's green and brown segmented body pattern blends with the leaves. When it is disturbed, the caterpillar withdraws its head into the first segment of its body and lifts the round red edge of the segment to face the threat. Two black spots on the red ring look like eyes. This adaptation mimics an animal with a big, fierce face.

Holes the caterpillar breathes through, called *spiracles*, create a pattern, too.

If the mean face doesn't work, the double tail can squirt nasty acid.

did you know?
GEESE FLY IN A "V" PATTERN TO SAVE ENERGY, BY REDUCING WIND RESISTANCE, AND TO KEEP TRACK OF THE OTHER GEESE IN THE FLOCK.

A cross-section cut reveals the beautiful pattern of chambers hidden inside a nautilus shell.

NAUTILUS SHELL ▶

The pearly nautilus is a cephalopod, an ancient group that includes octopuses. Nautiluses build their shells as they grow, making each chamber slightly larger than the one before it. Nautiluses use these chambers to regulate the amount of gases and water needed to keep them afloat and upright. The animal lives in the outermost chamber of a spiral.

As it grows, the nautilus seals off the cramped rear part of its living quarters and extends the front edge.

BIOMIMETICS

A gecko runs across the ceiling. Tiny hairs on its feet help it stick tight without falling. Robots cling to the inside of a volcano, climbing and peeling their feet just like geckos. Are these similarities a coincidence? Not at all. Nature has fantastic designs, and engineers often copy them to make their own inventions better. The mimicry of ideas from nature is called *biomimetics*. Many Olympic swimmers wear suits with bumps like those on a shark to help them move quickly through the water. A swarm of ants seems to wander in all directions, yet there aren't many collisions, so traffic engineers have tracked the ups, downs, ins, and outs of an ant colony to improve the flow of car and airplane traffic. Organisms have had many generations to develop great designs and adaptations—and people are realizing that we can learn from them.

did you know? ONE COMPANY IMPROVED THE FUEL EFFICIENCY OF A CAR BY MODELING ITS DESIGN ON THE SHAPE OF A FISH.

HITCHING A RIDE ▼

In 1948, Swiss engineer George de Mestral walked through the woods with his dog. Afterward, he pulled hundreds of burs from his pants and from his dog's fur. These sticky burs contained the seeds for the burdock plant. On each bur, he noticed a star of spines with a tiny hook at the end. The hooks grabbed onto a passing animal and the seeds inside hitched a ride to spread the plant through the woods and fields.

The nylon hooks on one side of a fastener grab loops of fabric on the other side and cling tightly until they are straightened by an upward force.

The hooks on the bur from the burdock plant helped it spread its seeds worldwide.

KEEPING THINGS TOGETHER ▶

Mestral used his observations to create a new fastener, which he named Velcro® (from the French words for "velvet" and "hook"). You know the familiar ripping sound of this fastener on your sneakers, jacket, and backpack, but it has more exotic uses, too. Astronauts use Velcro in their helmets and to keep tools in place in zero gravity. Velcro even held a human heart together during the first artificial heart surgery.

ON THE WINGS OF A WHALE? ▶

A humpback whale is as heavy as a loaded tractor-trailer truck, but it swoops through the water gracefully, propelled by its broad fins. An airliner must move through the atmosphere in the same way. Marine scientists study the shape of the whale's fins, and aeronautical engineers are designing new airplane wings to mimic them. The next time you have a smooth flight, you can thank a whale.

Think of a jet flying through the air as a model of a whale gliding through the water.

Bumps on the front edges of a whale's fin make the fin more efficient. This shape is also mimicked by wind turbines and helicopter blades.

FERNS

In the world of plants, ferns may appear to be fragile and delicate, but their looks are deceiving. Although most ferns are found in warm, tropical areas, ferns are hardy, widespread, and adaptable plants. They are found all over the world and can grow in forests, on the surface of ponds, and even in rocks. Ferns are the largest group of seedless, vascular (veined) plants on Earth. There are at least 11,000 known fern species and probably hundreds more that have yet to be discovered. Unlike flowering plants, ferns do not use seeds to reproduce. Instead, they release tiny structures called *spores* that are scattered by the wind. These spores develop into heart-shaped plants called *prothallia*, which are so small they are hard to notice in the wild. Sperm and eggs are produced on the prothallia. They combine to form embryos that grow out of the prothallia into new ferns.

From fern fossils we know that ferns belong to an ancient group of plants that evolved between 385 and 359 million years ago.

TREE FERNS ▶

Super-sized ferns are common in the world's tropical forests. They are called tree ferns and, even though they are not true trees, they can grow to be 80 feet (about 24 m) tall. Tree ferns don't have bark. Instead, their trunks are made of vertically growing rhizomes, or stems, covered by a sheath of durable roots.

FIDDLEHEADS ▼

Fern buds and leaves develop in a tightly coiled structure that looks like, and is called, a *fiddlehead*. As the young fern leaves mature, the fiddleheads begin to slowly unfurl. Fiddleheads are one of the most recognizable of a fern's features, although not all ferns have them. Some people consider fiddleheads a delicacy and collect, cook, and eat them each spring.

▲ THE COMMON WOODLAND FERN

One unpopular but common fern species is a woodland fern called *bracken*. It is found in temperate and tropical climates worldwide. Bracken is one of the first plants to colonize open ground, invading fields and pastureland.

A structure called an *indusium* protects clusters of the spore-producing structures called *sporangia*. Sporangia are found on the undersides of fern fronds.

The leaves of tree ferns can be up to 13 feet (almost 4 m) long.

did you know?................

NATIVE AMERICANS USED TO SNACK ON THE SWEET TASTING LICORICE FERN. THEY USED THE LADY FERN TO MAKE TEA.

FLOWERS

Flowers are loved all over the world for their sweet smells and beautiful colors, but sometimes they are deadly. Certain flowers are so poisonous that children have died from eating them or from drinking vase water. Other flowers are edible and nutritious. With over 400,000 species, flowers have a variety of sizes, shapes, colors, and smells. Some, such as the dandelion, are familiar because they grow in so many places. Others, such as the giant *Rafflesia*, are rare and have unusual characteristics. You might think of a flower as a plant with colorful blossoms, but to a scientist, a flower is only the reproductive part of a plant: the part that makes seeds. Strong smells and bright petals attract insects. The reward for visiting is a sugary liquid food called *nectar*. Pollen sticks to insects when they eat, and they transport the pollen between flowers. Parts of the pollen then fertilize eggs, which develop into seeds.

Fruit containing seed is carried into the air.

FLYING FRUIT ▶

Dandelion seeds have feathery parachutes to help them fly far from their parent plant. A dandelion is made up of many small flowers, called florets. Each floret develops a single fruit. The fruits form inside the closed-up seed head, after the yellow petals have withered away. When the weather is dry, the seed head opens, revealing a ball of parachutes. The slightest breeze lifts the parachutes into the air.

The parachute remains attached to the seed head until a gust of wind picks it up.

The open seed head reveals fully formed fruits, each attached to a tiny parachute.

The flowerhead opens daily. Insects get nectar from the flower but do not pollinate it.

Seeds develop inside the closed flowerhead.

did you know? HONEY BEES VISIT ABOUT TWO MILLION FLOWERS TO MAKE ONE POUND (0.5 KG) OF HONEY.

RAFFLESIA ▼

Not all flowers have the sweet smell you might expect. The world's largest flower is a species of *Rafflesia*. This enormous flower smells like rotting meat. The smell is repulsive to most people (although it doesn't seem to be bothering these kids!), but it attracts insects that pollinate the flower. It has no stem, leaves, or roots. Instead, it lives as a parasite. It obtains nutrients from thin strands of tissue that extend into nearby vines. *Rafflesia* grows only in the rain forests of Southeast Asia.

❶ GIANT PETALS

A fully grown *Rafflesia* is about 3 feet (1 meter) wide. The bud takes almost a year to mature. It then opens into five thick, fleshy petals which last only about a week before decomposing. This short time available for pollination is one reason the plant is rare.

❷ PROTECTIVE SPIKES

Hard spikes extend from the center region of the *Rafflesia*. In a female flower, these spikes protect thousands of eggs.

❸ RIM OF THE CUP

The rim of tissue around the central cup of a *Rafflesia* tends to keep the putrid smell close to the flower. The strong smell attracts more insects and increases the chance of pollination.

PLANT TRICKS

You might think all plants rely only on photosynthesis or absorb all their nutrients from the soil. Think again! Plants live in a wide variety of biomes and habitats such as forest, desert, and marsh. They have evolved adaptations, characteristics that help them survive in stressful environments. Some plant adaptations are especially unique. Places like rain forests and many swamps, for example, have very little fertile soil. Plants that inhabit these places cannot depend on the soil to fulfill all their nutritional needs. They must find other ways to survive. Some live on nutrients left in the dead matter of other plants. Others actually trap, kill, and digest insects. Because some plants like Indian pipe (white waxy plants) lack chlorophyll, they cannot make their own food to live. Instead, they have a parasitic relationship with fungi, where the plant's roots tap into the fungi and take the nutrients they need.

VENUS' FLYTRAP ▼

The Venus' flytrap lives in certain boggy areas of North and South Carolina. Because the soil cannot provide all its nutrients, the Venus' flytrap evolved as a carnivorous plant. Carnivorous plants are adapted to attract, capture, digest, and absorb insects and other tiny animals. The flytrap's jawlike leaves secrete sweet-smelling nectar. When an unsuspecting insect tickles the trigger hairs, the leaves snap shut. Digestive fluids break down the insect's soft parts, and the leaves absorb the nutrients. The plant releases the exoskeleton days later.

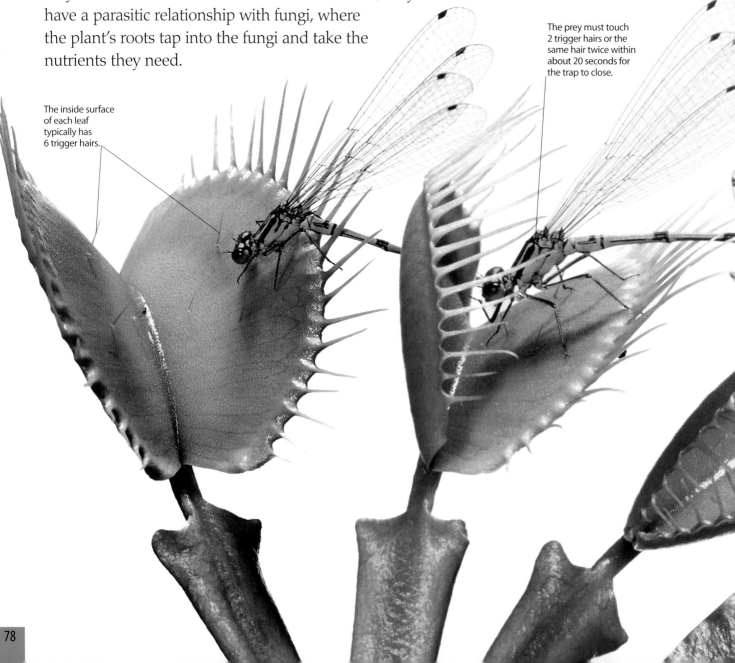

The prey must touch 2 trigger hairs or the same hair twice within about 20 seconds for the trap to close.

The inside surface of each leaf typically has 6 trigger hairs.

When the pitcher is fully grown, the lid opens and the trap is set.

Certain fly and mosquito larvae are known to inhabit some pitcher plant species.

Some pitchers grow at the end of a tendril that hangs from a flat leaf.

In some species, bacteria living in the pitcher excrete the enzymes that digest prey for the plant.

The trap does not close entirely at first, enabling very small insects that would not provide enough nutrients to escape.

After 3 to 5 meals, the leaves stop trapping prey and simply photosynthesize for several months before dropping off the plant.

◄ PITCHER PLANT

The various species of carnivorous pitcher plants evolved to have a leaf shaped like a tall pitcher or cup. The leaf partially fills with a fragrant liquid that attracts prey. Once inside the pitcher, the prey slides down the slippery lower walls of the leaf. Downward pointing hairs above the slippery sides prevent the prey from escaping. After the prey drowns in the liquid, acid from digested prey or enzymes produced by bacteria digest the prey and the leaf absorbs the nutrients.

did you know?...
BROMELIADS ARE LIKE SMALL ECOSYSTEMS. BIRDS, TREE FROGS, TINY CRABS, AND OTHER CREATURES MIGHT SPEND THEIR ENTIRE LIVES INSIDE THEM.

AIR PLANTS ▲

Bromeliads, also called *air plants,* do not need soil. Instead they use their roots to cling to a host plant. Bromeliads get all the moisture and nutrients they need from the air and from leaves and debris that fall into them. Their leaves form a tight spiral, capturing water from dew or a rainstorm. They also are home to insects that excrete wastes full of nutrients.

BIOFUELS

Fields of flax, sunflowers, and corn—do they make you think of cars? These plants all play a part in making liquid fuels, called *biofuels*, that can power cars, tractors, buses, and more. Biofuels are made from biomass, which is matter from living things, especially plants. One advantage of using more biofuel is that we use less fossil fuel, such as gasoline. Another benefit is that not only are plants renewable, but the carbon dioxide (CO_2) they release when they burn is balanced by the CO_2 they take in when they grow. Some negatives are that gases are given off during the production and transport of biofuels. The process of clearing land in order to turn it into cropland also generates greenhouse gases. Scientists are hoping that on balance the total output of CO_2 is less, but this is still being debated.

CORN IN YOUR GAS TANK? ▶

One biofuel is ethanol, an alcohol often made from high-starch grains such as corn. Ethanol can be used on its own as a fuel or blended with gasoline. Henry Ford designed the Model T to run on ethanol. Scientists are working on ways to make ethanol from parts of plants that are not food, such as husks, stems, and leaves.

The use of corn is controversial—should the land be used for food or fuel?

did you know?
THE DIESEL ENGINE DEMONSTRATED AT THE 1900 WORLD'S FAIR RAN ON PEANUT OIL.

PLANT POWER ▶

These lavender flax and yellow canola plants yield oil that may soon help make biodiesel, another biofuel. Biodiesel is made by mixing alcohol with vegetable oils. Even recycled cooking grease or animal fat can be used. Biodiesel can be added to gasoline or used on its own in diesel engines.

MORE THAN JUST A PRETTY FACE ▲

Though it can be a bit costly, sunflower oil is being looked at as a source for both ethanol and biodiesel fuel. Some farmers make biodiesel fuel from oil extracted from the sunflowers they grow. They mix the oil with lye and alcohol to create biodiesel to fuel their trucks and tractors.

NOT JUST A FAST CAR ▼

This British car, the Vauxhall Astra, runs on 100 percent ethanol. In fact, many racing cars use ethanol because it combusts, or burns, more completely in the engine, giving great performance and reducing emissions and smog. In 2007, the Indy Racing League®, home of the Indianapolis 500®, began using 100 percent ethanol as its official race fuel.

Yellow canola plants   Flax

SEED BANK

In a vault on the Norwegian island of Spitsbergen near the North Pole sits a valuable treasure. It's not gold or diamonds, but seeds! The future of the world's food supply (and all other products we get from plants) rests with the seeds stored here in the Svalbard Global Seed Vault and in the more than 1,400 seed banks around the world. Each plant species may contain many varieties. For example, there are more than 100,000 different varieties of rice! Genetic differences within a species may mean differences in resistance to disease or in the ability to grow in different climates. At the same time, of the thousands of plant species that have historically appeared in peoples' diets, fewer than 150 are grown today. Plant species threatened with extinction due to natural disaster, disease, war, and climate change are helped by seed banks that maintain the diverse gene pools needed to feed the world.

DOOMSDAY VAULT ▶

This unusual-looking structure is the entrance to the Svalbard Global Seed Vault. Nicknamed the "Doomsday Vault" and "Noah's Ark," it was built by the Norwegian government to provide a secure location for seed banks around the world to store duplicate collections. Its 3 storage chambers can hold a total of 4.5 million seed samples, each sample containing an average of 500 seeds.

INTERNATIONAL CENTER OF ▲ TROPICAL AGRICULTURE

The International Center of Tropical Agriculture (CIAT) in Colombia holds the world's largest and most varied collection of beans, cassava, and tropical grasses from Latin America, Asia, Africa, and the Middle East. For more than 40 years, scientists and farmers have used these materials for research and agricultural purposes. The CIAT has already sent duplicates to the Svalbard vault for safekeeping.

▲ DEEP IN A FROZEN MOUNTAIN

The remote island location of the Svalbard Vault was selected in part because of its climate and geology. The deep underground location of the storage chambers in frozen ground will keep the vault's temperature cool enough to protect the seeds in case of a power failure. The facility's elevation—about 427 feet (130 meters) above sea level— will protect the seeds from any rise in sea level resulting from global warming.

did you know?

IN 2005, A 2,000-YEAR-OLD SEED FROM A NOW-EXTINCT SPECIES OF DATE PALM WAS SUCCESSFULLY SPROUTED, THE OLDEST KNOWN SUCH SPROUTING!

ANIMAL BODIES

Porcupine spines, elephant trunks, and delicate butterfly wings. You can recognize many animals by the special structures, or parts, of their bodies. But these structures do more than just show who's who. They also help the animal survive. For example, an elephant's trunk is long, strong, and flexible, making it useful for grabbing food, collecting water, and interacting with other elephants. Animals' bodies are loaded with parts that help them stay alive. Take eyes, for example. Beavers have an inner eyelid that they can see through under water. A similar eyelid keeps sand out of camels' eyes. Frogs' eyeballs can sink back into their skulls to help push large prey down their throats.

A porcupine's spines are made out of the same protein that makes up hair and fingernails.

LITTLE ARMORED ONE ▼

Armadillos are the only mammals alive with such a hard, leathery shell. *Armadillo*, in fact, means "little armored one" in Spanish. This nine-banded armadillo is an insect-eating mammal about the size of a small cat. It lives in the southeastern United States. The armadillo's shell offers good protection from its enemies. When not hunting for food, the armadillo spends most of its time sleeping in an underground burrow.

Mammals, such as porcupines and armadillos, have hair. This important structure helps them stay warm.

The bands on a nine-banded armadillo are joints that let the animal bend its body. Otherwise, its armor is not very flexible!

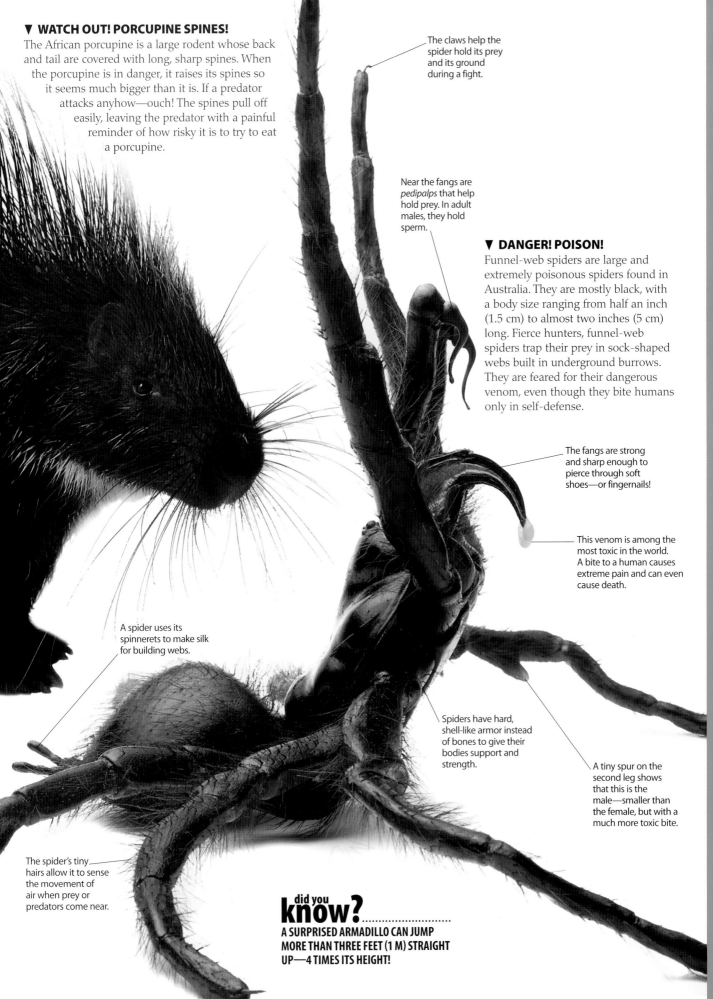

▼ WATCH OUT! PORCUPINE SPINES!

The African porcupine is a large rodent whose back and tail are covered with long, sharp spines. When the porcupine is in danger, it raises its spines so it seems much bigger than it is. If a predator attacks anyhow—ouch! The spines pull off easily, leaving the predator with a painful reminder of how risky it is to try to eat a porcupine.

The claws help the spider hold its prey and its ground during a fight.

Near the fangs are *pedipalps* that help hold prey. In adult males, they hold sperm.

▼ DANGER! POISON!

Funnel-web spiders are large and extremely poisonous spiders found in Australia. They are mostly black, with a body size ranging from half an inch (1.5 cm) to almost two inches (5 cm) long. Fierce hunters, funnel-web spiders trap their prey in sock-shaped webs built in underground burrows. They are feared for their dangerous venom, even though they bite humans only in self-defense.

The fangs are strong and sharp enough to pierce through soft shoes—or fingernails!

This venom is among the most toxic in the world. A bite to a human causes extreme pain and can even cause death.

A spider uses its spinnerets to make silk for building webs.

Spiders have hard, shell-like armor instead of bones to give their bodies support and strength.

A tiny spur on the second leg shows that this is the male—smaller than the female, but with a much more toxic bite.

The spider's tiny hairs allow it to sense the movement of air when prey or predators come near.

did you
know?..........................
A SURPRISED ARMADILLO CAN JUMP
MORE THAN THREE FEET (1 M) STRAIGHT
UP—4 TIMES ITS HEIGHT!

NEW BODY PARTS

Our bodies can grow new blood cells, hair, nails, and skin. They can repair minor bone injuries and regrow liver tissue. But they cannot regrow entire limbs the way some animals can. We can cut flatworms and sea sponges into pieces, and each piece will grow into a complete animal. More complex animals like crabs and lobsters can regrow claws, eyes, and legs. Regenerating body parts involves unspecialized cells, called *stem cells,* that can grow into any type of cell. Scientists are researching these animals that can regenerate body parts in hopes they can one day duplicate limb regeneration in humans.

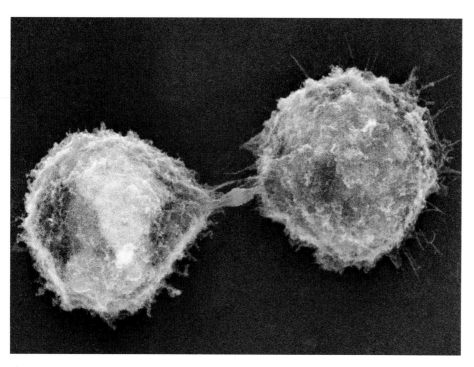

SEA STARS ►

Most sea star species can regrow an arm if the central disc (the center of the body) is undamaged. A few species can regrow an entire body from a single arm. Scientists have found a DNA structure (gene) in gray sand star larvae that is similar to one in human embryos. The gene is involved in embryonic development and wound repair. Scientists believe they can use gray sand star larvae as a model to research wound healing and tissue regrowth in humans.

▲ HOW DO BODY PARTS REGENERATE?

Stem cells renew themselves by cell division, giving rise to two identical daughter cells. Then, through a process called *specialization,* they can become any type of cell—skin, muscle, nerve, and other tissues. Specialized cells can revert to stem cells in animals that regenerate body parts. Cell-to-cell communication gives the stem cells instructions, such as what type of cells to become and where they belong.

When a lizard tail breaks off, it separates on a special plane in the middle of a vertebra.

THE LIZARD'S ESCAPE STRATEGY ▲

Many lizard species can voluntarily break off their tails to get away from attacking predators. Afterward, the lizard can regenerate skin, muscle, fat, cartilage, and neural tissues to regrow a new tail. The new tail, however, is not as long or strong as the original as it has a cartilage rod rather than vertebrae.

This sea star is regenerating two arms.

ECHOLOCATION

Most animals use vision to find prey and travel through their world. But some animals use their ears to "see." These animals emit sounds (often clicks or squeaks) that travel in waves through air or water. When a sound wave encounters a solid object, it bounces back. By listening to this echo, animals can locate prey or obstacles. The echoes give information about an object's size, shape, distance, traveling speed, and direction of movement! This process is called *echolocation*. Some bats, birds, shrews, and marine animals use echolocation. The time it takes for an echo to return shows how far away prey is. The loudness of an echo can indicate the prey's size, distance, and even texture. Dolphins can detect prey from hundreds of yards or meters away, and bats can tell whether or not a moth is fuzzy. When bats begin a hunt, they may send out one sound per second. As they get close to their prey, however, they may emit 200 or more sounds per second.

▼ THE ECHO OF A WORM

Pygmy shrews weigh less than a penny and can fit on your thumbnail. They must eat every 15 to 30 minutes, 24 hours a day. When hunting prey, pygmy shrews make a rapid yawning motion. They aren't sleepy. They are emitting ultrasonic pulses of sound to echolocate prey, such as this worm. Pygmy shrews also use echolocation to find their way through piles of leaves or to navigate in underground tunnels.

Shrews have tiny eyes and must rely on touch (whiskers) and sound (echolocation) to locate prey.

Shrews have two sets of whiskers, which they can twitch 20 times per second to touch and determine characteristics of prey.

Shrews and moles eat insects and earthworms. Earthworms avoid sunlight, and instead come to the surface to feed at night.

DINING LATE ▶

Bats have good eyesight, but they are nocturnal mammals. Many bats use echolocation to navigate and find food in the dark. Different types of bats use different echolocation frequencies. Scientists can record the sounds to identify bat types. Moths have evolved strategies to defend themselves from echolocating bats. Some have fuzzy wings that muffle the echo. Foul-tasting tiger moths send clicking noises out to bats, and bats avoid them.

did you know?.........................

USING ECHOLOCATION, A BAT TRAVELING AT HIGH SPEEDS CAN DETECT AN OBJECT THE WIDTH OF A HUMAN HAIR!

Bats emit bursts of high-pitched sounds to locate prey.

Echoes bounce back from a nearby moth.

It takes a bat less than half a second to capture a moth after detecting it.

INSECTS

With more than 1,000,000 identified species, insects are the most abundant animals on Earth. Scientists estimate that millions more species of insects have yet to be identified. Insects have adapted to jungles, deserts, the arctic tundra, and even hot springs. They are remarkably diverse in their size, appearance, and behavior. Some can swim while others fly. Many jump, play dead, and even sing. Insects range in length from less than a millimeter (fairy fly) to more than a foot (giant walking sticks of Malaysia).

BUTTERFLIES ▶

Butterflies and moths are in the insect order Lepidoptera. Butterflies have four wings covered with colorful scales and a coiled tongue that they use to sip nectar. They inhabit every continent except Antarctica. There are at least 14,000 known species of butterfly and probably thousands more waiting to be discovered.

MOTHS ▲

Unlike butterflies, moths are mostly nocturnal, and their antennae are feathery rather than club-shaped. There are at least 100,000 known species, the largest of which has a 1-foot (about 30 cm) wingspan! A moth's wings lie flat against its back when at rest; a resting butterfly's wings point straight up.

Great eggfly butterfly pupa

Red admiral butterfly

Butterfly eggs

Swallowtail butterfly

Pink sallow moth

Oleander hawk moth

Mottled green moth

Caterpillar

Oak silk moth pupa

Brimstone moth

Polyphemus moth

Dragonfly

Housefly

Blowfly

Cranefly

Maggots

Hoverfly

BEES AND BEETLES ▼

Bees belong to the order Hymenoptera. Bees are nectar-eating insects responsible for pollinating flowers, trees, and many of our food crops. Scientists estimate that there are about 20,000 species of bee. Beetles are in the order Coleoptera, the largest insect order. They are perhaps the most diverse and successful animals on Earth, with more than 350,000 known species. Beetles are characterized by their hardened forewings, called *elytra*, that protect the hindwings.

Wood ant

Wasp

Honeycomb

Bee

did you
know?........................
MORE THAN 75 PERCENT OF ALL KNOWN ANIMAL SPECIES ARE INSECTS.

Jewel wasp

Cockroach

Walking stick insect eggs

Chafer beetle

Dung beetle

Stag beetle

Hissing cockroach

Fighting stag beetles

Hercules beetle

Mealworm

Ladybug beetle

Jeweled frog beetle

Leaf insect

Mealworm

Stag beetle

Frog beetle

Giant walking stick

Housefly

Red-spotted longhorn beetle

SPIDERS

It's hard to think of a creepy-crawly spider as an animal, isn't it? Well, it is. Spiders belong to the planet's most numerous and diverse group of animals: the invertebrates. Scientists have identified about 1.7 million invertebrate species, almost 40,000 of which are spiders. Like all invertebrates, spiders lack a backbone. Instead, they have an exoskeleton—a hard outer shell that protects them from predators, shelters their internal organs, and regulates water loss. Like many invertebrates, spiders are bilaterally symmetrical. That means a spider has a distinct front and back, and when you draw a line down its middle, the left and right sides mirror each other. Spiders are not insects; they are arachnids. Arachnids have two distinct body sections: the head section, called the *cephalothorax*, and the abdomen, or body section. Spiders also have four pairs of walking legs, two fangs called the *chelicerae*, and two pedipalps—leglike structures that function as sense organs.

The cephalothorax houses the stomach, brain, and central nervous system.

Pedipalp

The carapace is a hard protective layer that covers the cephalothorax.

The abdomen contains the heart, intestines, and reproductive organs.

Silk is produced by silk glands and spun with structures called *spinnerets*, which are located beneath the abdomen.

▼ SPINY-BACKED ORB WEAVER

Although spiders all share the same basic body plan, some are more ornate than others. The remarkable, tiny (2–10 mm) spiny-backed orb weaver is found throughout the southern United States and South and Central America. The female has 6 rosy-red spines protruding from her kite-shaped abdomen. Male spiny backs are smaller and less colorful than the females, and they don't have spines, just 4–5, small, abdominal humps.

The spines may serve as defensive structures that discourage predators from making a meal of the little spider.

did you know?..............
TARANTULAS CAN LIVE FOR MORE THAN 20 YEARS! THAT'S LONGER THAN MOST DOGS LIVE.

▼ JUMPING SPIDERS

With big eyes, fuzzy bodies, and flat faces, jumping spiders are pretty cute, as spiders go. Ranging in size from 2 to 22 mm, members of this family do not build webs to catch prey. Instead, they use their extraordinary jumping ability and keen binocular eyesight to pounce on prey with marvelous accuracy. Scientists believe that jumping spiders have the best eyesight of any spider. That's because they have rows of eyes on the front, top, and sides of their heads, so they can see close-up, wide-angle, and distance. Like most spiders, jumping spiders have four pairs of eyes, with each pair used for a different type of vision.

Some spiders can spin up to eight different types of silk, including egg sac silk.

Some spiders guard their egg sacs and care for the young. Others abandon them after the sacs are made.

▲ EGG SAC

Female spiders lay hundreds to thousands of eggs in silken sacs. These egg sacs keep the eggs warm and protect them from predators. Some spiders attach the sacs to the bottoms of leaves or stones to hide them. Other spiders carry the sacs in their jaws or attached to their spinnerets until the spiderlings hatch.

EXOSKELETON

Clams, spiders, lobsters, and snails all have their own body armor, called an *exoskeleton*. These tough coverings act as protection, keeping the outside out and the inside in. They also support the creature living inside. Like our skin, these coverings help protect the animal from drying out. Unlike skin, exoskeletons do not automatically increase in size when the animal inside grows. Some animals, such as clams, are able to add onto their shells layer by layer. Insects and crustaceans such as crabs, with more complicated body shapes, have a different solution. These creatures molt, or shed, their exoskeleton regularly as they grow. Some even reabsorb or eat their old outer covering to keep the valuable proteins inside.

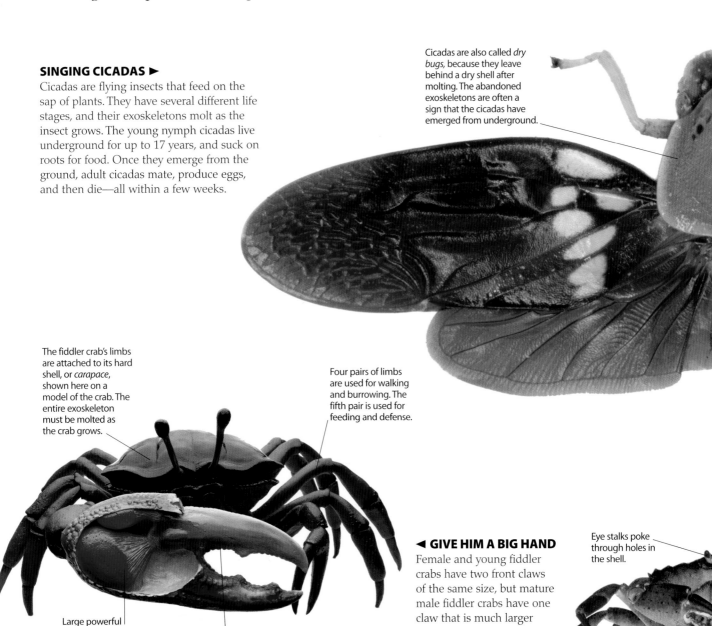

SINGING CICADAS ▶

Cicadas are flying insects that feed on the sap of plants. They have several different life stages, and their exoskeletons molt as the insect grows. The young nymph cicadas live underground for up to 17 years, and suck on roots for food. Once they emerge from the ground, adult cicadas mate, produce eggs, and then die—all within a few weeks.

Cicadas are also called *dry bugs*, because they leave behind a dry shell after molting. The abandoned exoskeletons are often a sign that the cicadas have emerged from underground.

The fiddler crab's limbs are attached to its hard shell, or *carapace*, shown here on a model of the crab. The entire exoskeleton must be molted as the crab grows.

Four pairs of limbs are used for walking and burrowing. The fifth pair is used for feeding and defense.

Large powerful muscles operate the top section of the claw, allowing it to open, close, and crush.

The male's single large claw is used to fight and to attract mates. It uses its small claw for feeding.

◀ GIVE HIM A BIG HAND

Female and young fiddler crabs have two front claws of the same size, but mature male fiddler crabs have one claw that is much larger than the other. This claw can be up to 65 percent of the crab's weight!

Eye stalks poke through holes in the shell.

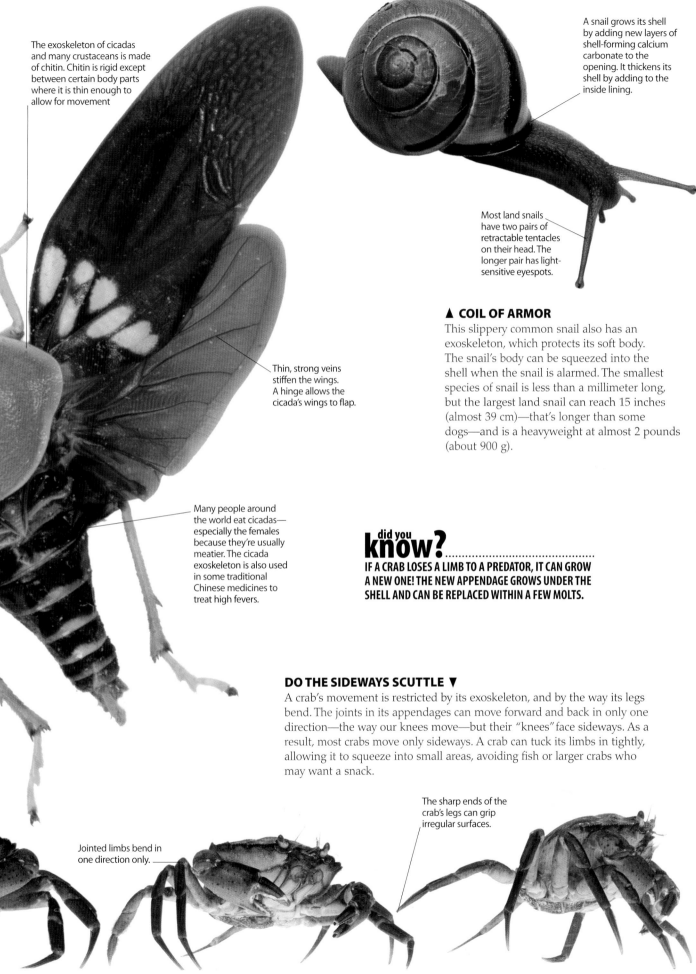

The exoskeleton of cicadas and many crustaceans is made of chitin. Chitin is rigid except between certain body parts where it is thin enough to allow for movement

A snail grows its shell by adding new layers of shell-forming calcium carbonate to the opening. It thickens its shell by adding to the inside lining.

Most land snails have two pairs of retractable tentacles on their head. The longer pair has light-sensitive eyespots.

Thin, strong veins stiffen the wings. A hinge allows the cicada's wings to flap.

Many people around the world eat cicadas—especially the females because they're usually meatier. The cicada exoskeleton is also used in some traditional Chinese medicines to treat high fevers.

▲ COIL OF ARMOR

This slippery common snail also has an exoskeleton, which protects its soft body. The snail's body can be squeezed into the shell when the snail is alarmed. The smallest species of snail is less than a millimeter long, but the largest land snail can reach 15 inches (almost 39 cm)—that's longer than some dogs—and is a heavyweight at almost 2 pounds (about 900 g).

did you
know?...........................
IF A CRAB LOSES A LIMB TO A PREDATOR, IT CAN GROW A NEW ONE! THE NEW APPENDAGE GROWS UNDER THE SHELL AND CAN BE REPLACED WITHIN A FEW MOLTS.

DO THE SIDEWAYS SCUTTLE ▼

A crab's movement is restricted by its exoskeleton, and by the way its legs bend. The joints in its appendages can move forward and back in only one direction—the way our knees move—but their "knees" face sideways. As a result, most crabs move only sideways. A crab can tuck its limbs in tightly, allowing it to squeeze into small areas, avoiding fish or larger crabs who may want a snack.

The sharp ends of the crab's legs can grip irregular surfaces.

Jointed limbs bend in one direction only.

SUPERCOOLING FROGS

Like all amphibians, frogs are ectotherms, meaning their body temperature depends upon their environment's temperature. Temperature is an important abiotic factor (nonliving part) of a frog's environment. How do some frogs that live in colder climates survive the freezing conditions of winter? The secret lies in their ability to freeze! Freezing typically damages cells—mostly due to ice formation—and can result in tissue death, such as frostbite, or death of the entire animal. However, some frogs have a secret anti-winter weapon. As soon as the frog's body detects ice, the frog's liver quickly produces large amounts of glucose or glycerol, which the circulatory system distributes to every part of its body. The glucose or glycerol is a cryoprotectant. It reduces ice formation in the cells and controls the freezing process. At the same time, water leaves the cells and moves into empty spaces in the frog's body, where it will freeze without damaging the cells. Eventually, the vital organs freeze and shut down, but the chemical interactions needed to keep the frog alive continue, even without oxygen!

SPRING PEEPER ▼

Amphibians have many ways of dealing with cold winters. Some species of toads dig deep holes. Newts hibernate underground in existing holes. Some aquatic species live in the water below the ice. Spring peepers, frogs from eastern North America that are about an inch (2.54 cm) long, can winter in less protected areas: under logs or up in trees.

GRAY TREEFROG ▶

Gray treefrogs inhabit eastern North America. They help control the insect population by eating mosquitos, flies, and gnats. When hibernating, they may burrow under fallen leaves or hide beneath loose bark. Both gray treefrog species—this eastern gray treefrog and Cope's gray treefrog—endure freezing in order to survive winters in the northern parts of their ranges.

The heart is the last organ to shut down during freezing and the first organ to start when thawing.

Ice growth typically starts in the frog's hind limbs and ends in the brain.

▲ WOOD FROG ON ICE

The temperature at which each of these frogs begins to freeze also plays an important role in helping them survive. A substance is supercooled if it remains liquid below the freezing point. The cryoprotectant in the frog's cells lowers the temperature at which the frogs will freeze. For each of these frogs, their bodies supercool only a few degrees before ice begins to form. For example, wood frogs supercool only to between 28°F and 26°F (−2°C and −3°C) before they start freezing. This reduces the possibility of ice growing too quickly, damaging the organism. Instead, the process of ice formation is carefully controlled.

BIRDS

This diverse and colorful crowd has about 9,000 species of all shapes and sizes. The 2.5-inch (about 6.4-cm) bee hummingbird weighs the same as a dime. The wandering albatross wins for longest wingspan: 11 feet (about 3.4 m). Ostriches can grow to 9 feet (2.7 m) tall. And because most birds can fly, they live almost everywhere, from deserts to the poles. Some species of birds do not travel far in their lifetime, while others migrate huge distances. Setting a nonstop flight record, a female bar-tailed godwit flew 7,257 miles (11,679 km) from Alaska to New Zealand.

Hummingbird

Zebrafinch

Blue tit

Tawny eagle

Emu

King penguin

Pekin duck

Peahen

TOO BIG TO FLY! ▶
Emus have long powerful legs, but very small wings. They cannot fly, but they do run fast—up to 31 miles (almost 50 km) per hour. Other flightless birds include the ostrich, rhea, cassowary, and kiwi.

FANCY FEATHERS ▼

The Indian peacock displays his amazing long tail feathers, called his *train*, to attract a mate. The peacock lifts his train to form a fan. Each feather has eye-like spots of brilliant blue, green, and orange. Some studies show that the more eye spots the peacock has, the more peahens he attracts.

Parakeet

Owl

Swift

TONGUE TWISTER ▶

A green woodpecker has a powerful beak. It also has a long, sensitive tongue that is stored curled up inside its skull. The tip of the tongue is armed with barbs, which keep the ants and grubs the bird eats from wriggling away.

Green woodpecker

Toucan

Flamingo

Pelican

Peacock

did you know?....................
BECAUSE THEY DON'T HAVE TEETH, BIRDS NEED A SECOND STOMACH, CALLED A *GIZZARD*, TO GRIND UP FOOD.

BIRDS CONTINUED

Birds use food and oxygen for "fuel," and both their frame and "engine" are built for fuel-efficiency. They have lightweight bones, many of which are hollow. For added strength, the bones have honeycombed supports inside. Birds digest food fast, and their large, fast-beating heart pumps nutrients quickly through their body. Their respiratory system is very efficient. Connected to their lungs are air sacs that tuck into spaces throughout the body. These sacs act like pumps, sucking in air and pushing it out of the body, keeping a continuous flow of fresh air to the lungs. The lungs themselves don't contract much. Even so, birds breathe rapidly during flight. Pigeons breathe more than 400 times per minute!

HUNTERS OF THE SKIES ▼

Falcons are streamlined, fast-flying birds of prey, called *raptors*, with long, narrow, pointed wings and long, narrow tails. They feed mainly on other birds and insects, and usually capture their prey in midair. Other raptors include owls, hawks, eagles, and vultures. Most of these birds have excellent eyesight and sharp, hooked bills for tearing at flesh. They are highly skilled fliers.

Most feathers are called *contour* feathers, including the wing and tail feathers, and are aerodynamic devices.

◀ FEATHERS FOR ALL OCCASIONS

Birds are the only animals that have feathers. There are many types—to insulate, streamline, and waterproof the bird, and to enable flight. Each bird has between 1,000 and 25,000 feathers, depending upon the species and size. The tundra swan has the most on record. Feathers are composed mostly of keratin, the same substance that horns, hooves, and fingernails are made from. Birds molt once or twice each year to replace old, broken, or worn-out feathers with new ones.

Fuzzy feathers, such as down, are for insulation. Colorful feathers are to attract mates.

Peacock down feather

Pheasant feathers

Peacock feather

The great hornbill, about 4 feet (about 1.2 m) tall, has a horny growth, called a casque, on top of its bill.

did you know?

SOME BIRDS' LARGE EYEBALLS ARE SO TIGHTLY PACKED INTO THEIR NARROW SKULL THAT THEY ALMOST TOUCH IN THE MIDDLE.

SHAPED FOR EATING ▲

Each type of bird has a bill or beak adapted for its particular diet. The great hornbill's sturdy beak can handle not only its favorite fruit, figs, but small animals. A strong cone-shaped bill helps finches and grosbeaks pick up and crack seeds. Insect-eaters such as warblers have thin, slender, pointed beaks. Hummingbirds have long, strawlike bills to sip nectar from flowers. A merganser grabs fish with the hooked point of its bill and holds it with the bill's jagged edges.

Some raptors can see a rabbit a mile (1.6 km) away.

In flight, the wing feathers closer to the body provide lift. The longer outermost feathers at the tip of the wing pull the bird forward.

Curved, sharp claws, called talons, help raptors catch and grip prey.

Tail feathers function as direction-changers and as brakes.

101

ALGAE

Some may be tiny—living as microscopic, single cells in soil, on rocks, or in water. Others may be tall—making up dense underwater forests of seaweed 100 feet (about 30 m) high. Whether tiny or tall, algae are important organisms! Algae are producers. Producers are living things that make their own food from carbon dioxide and water, using the energy from sunlight. The oxygen made by algae helps living things all over the world survive. And, as producers, algae form the base of many aquatic food chains. They are especially important in ocean food chains. There, the algae living in the sunlit upper waters are food to countless other organisms. And those organisms are food for even larger organisms, and so on up the food chain. In this way, the energy of sunlight is transferred from one living thing to another throughout the ocean.

WALLS OF GLASS ▶

Diatoms may seem like tiny glass ornaments. In fact, they are microscopic algae that have cell walls made of silica, the main substance that makes up sand and glass. Most diatoms live as single cells in oceans, lakes, and soils. It is estimated that diatoms carry out more than a quarter of the world's photosynthesis.

Ocean diatom found near the island of Oahu, Hawaii

Diatoms can have a wide variety of shapes, including long rods, round discs, fat cylinders, and stars, such as this one.

did you know? GIANT KELPS ARE MANY-CELLED ALGAE THAT CAN GROW TO A LENGTH OF 200 FEET (60 M).

SUNLIGHT HARVESTERS ▼

Algae, such as this single-celled desmid, are like tiny solar panels. The broad, flat shapes of algae can maximize the amount of sunlight that they take in. They use sunlight to produce sugars and oxygen through the process of photosynthesis. Photosynthesis takes place in cell structures called *chloroplasts*. A pigment called *chlorophyll* found in chloroplasts gives many algae and plants their vivid green color.

Freshwater desmid

The nucleus of a desmid is located in the very center of the cell.

A desmid reproduces by splitting along this central line. A new half grows onto each of the original half cells.

◄ VISIBLE FROM SPACE

When conditions are right, algae can multiply quickly in a very short period of time. These quick population spurts are called *algal blooms*. Algae called *coccolithophores* live in ocean surface waters, where they can get enough light to carry out photosynthesis. They can sometimes form algal blooms so large that they can be seen from space.

From the space shuttle *Discovery*, trillions of single-celled algae look like a turquoise cloud swirling in the Atlantic Ocean.

OCTOPUS

Although octopuses spend a lot of time lounging or strolling along the ocean floor using the suckers on their arms, they can move fast if they are hunting for food or escaping danger. Octopuses propel themselves by repeatedly taking in and squirting water out of a tube near their eyes, called a *funnel*. This maneuvering requires a well-developed circulatory system. Octopuses are one of the few invertebrates with a closed circulatory system, which means their blood is contained and transported inside the blood vessels. This allows blood, dissolved oxygen, and nutrients to travel quickly through the body. Octopuses power their circulation with three hearts: one large heart that pumps blood to the body and two smaller hearts that pump blood to the gills, which are responsible for respiration.

The blue-ringed octopus only displays its vivid blue rings when threatened. The bright pattern warns predators to steer clear of this venomous octopus.

OCTOPUS ANATOMY ▼

Octopuses belong to the order Cephalopoda, which means "head foot." Their eight boneless, muscular arms branch out of their heads. Each arm has two rows of super-strong suckers. The big sack behind the octopus's eyes, which looks like a giant forehead, is called its *mantle*. It contains most of the octopus's organs. The head includes a mouth, with a hard beak, as well as the eyes and brain.

Mantle

Salivary (venom) gland

Stomach

Digestive gland

Heart (one of three)

Brain

Eye

Gills

Funnel

Arm

Mouth

Sucker

◄ BLUE-RINGED OCTOPUS

Blue-ringed octopuses, which live off the Australian coast, produce highly toxic venom for which no known antidote exists. In just minutes, the venom can kill an adult human! The toxin is produced by bacteria that live in the octopus's salivary glands. Despite their deadly venom, blue-ringed octopuses are shy creatures that weigh less than a golf ball. They are only aggressive toward humans when provoked.

OCTOPUS CAMOUFLAGE ▼

Octopuses are nature's quick-change artists. In the blink of an eye, they can change their skin's color, pattern, and texture by triggering color-changing pigment cells in their skin, called *chromatophores*. Octopuses use this ability to blend into their environment or to startle would-be attackers. When in danger, octopuses will also shoot ink from an ink gland, giving them time to escape.

did you know?
OCTOPUSES ARE THOUGHT TO BE THE SMARTEST INVERTEBRATES. IN CAPTIVITY, THEY HAVE BEEN TAUGHT TO UNSCREW THE LID ON A JAR OF SHRIMP.

SHARKS

Sharks were here long before dinosaurs, and they haven't changed much in the last 100 million years. Sharks are "apex" predators, meaning they are at the top of the food chain. They help maintain the balance of their ocean environment by keeping the populations of prey animals from becoming too big. Often, they prey on sick and dying animals of many species. These include whales and seals, which have few predators because of their size. The shark population tends to increase or decrease in size as the size of its prey populations expand or shrink. However, the relationship between shark populations and their prey may be thrown out of balance due to overfishing by an unnatural predator of sharks—human beings.

The spotted catshark pup emerges from the egg capsule.

The newly hatched pup is about 3.5 to 4 inches (9–10 cm) long.

HATCHING ▲

Most sharks give birth to live young. Some sharks—mostly small ones like the spotted catshark—lay eggs and attach them to rocks or seaweed. The eggs have thick cases that protect the embryos from predators. A yolk sac within the egg capsule provides nourishment as the embryo grows. The newly hatched shark is called a *pup*.

The egg capsule of the catshark and some other sharks is called a *mermaid's purse*.

KEEN SENSES ▶

Sharks use some common senses such as vision, hearing, and smell. Other senses, such as electroreception and the lateral line—pores along the flank and back—are used mostly by aquatic animals. Electroreception makes sharks incredibly sensitive to electrical fields. This sensitivity allows sharks, like the oceanic whitetip shark shown here, to find hidden prey that can't be detected with their other senses. The lateral line system enables sharks to sense waves of pressure or disturbances in the water.

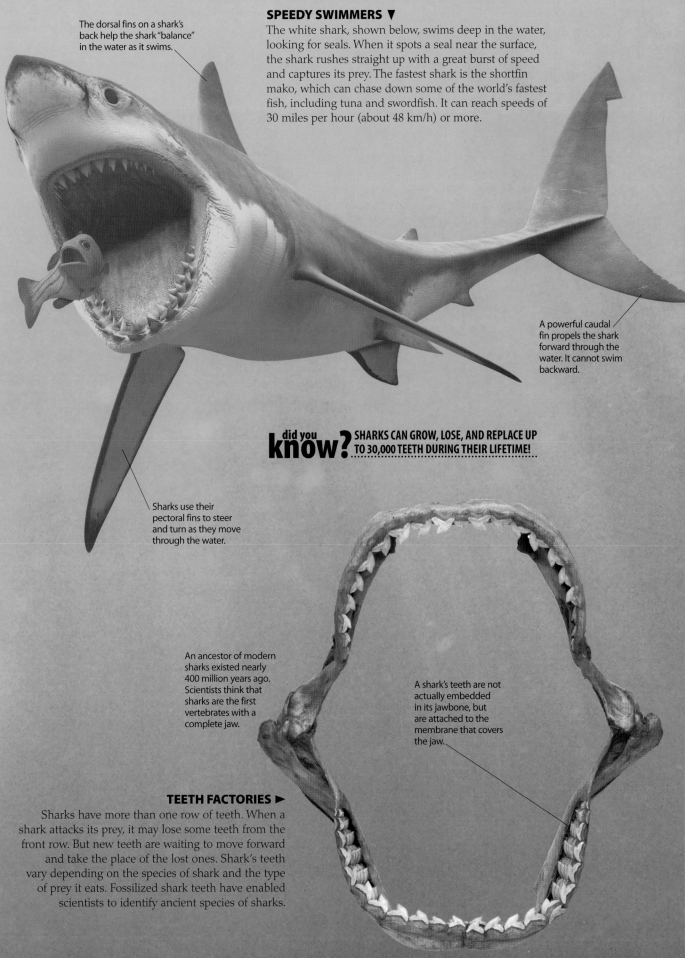

The dorsal fins on a shark's back help the shark "balance" in the water as it swims.

SPEEDY SWIMMERS ▼

The white shark, shown below, swims deep in the water, looking for seals. When it spots a seal near the surface, the shark rushes straight up with a great burst of speed and captures its prey. The fastest shark is the shortfin mako, which can chase down some of the world's fastest fish, including tuna and swordfish. It can reach speeds of 30 miles per hour (about 48 km/h) or more.

A powerful caudal fin propels the shark forward through the water. It cannot swim backward.

did you know? SHARKS CAN GROW, LOSE, AND REPLACE UP TO 30,000 TEETH DURING THEIR LIFETIME!

Sharks use their pectoral fins to steer and turn as they move through the water.

An ancestor of modern sharks existed nearly 400 million years ago. Scientists think that sharks are the first vertebrates with a complete jaw.

A shark's teeth are not actually embedded in its jawbone, but are attached to the membrane that covers the jaw.

TEETH FACTORIES ▶

Sharks have more than one row of teeth. When a shark attacks its prey, it may lose some teeth from the front row. But new teeth are waiting to move forward and take the place of the lost ones. Shark's teeth vary depending on the species of shark and the type of prey it eats. Fossilized shark teeth have enabled scientists to identify ancient species of sharks.

WHALES

Whales are huge mammals that have adapted to life in the water. The blue whale is the biggest animal living on Earth. The largest blue whale on record weighed about 300,000 pounds (about 136,000 kg) and measured more than 108 feet (about 33 m) in length. To maintain their huge sizes, whales eat a lot. The diets of different whale species vary quite a bit. Some species of whales have teeth—for example, orcas or "killer whales." They eat different kinds of fish but also hunt seals, sea lions, and even sharks! Other species, such as the blue and the humpback whales, have bony plates called *baleen* in the upper jaw. Baleen whales eat millions of zooplankton and tiny fish every day. They gulp great amounts of water and strain the food using their baleen. A whale's body is covered by a thick layer of fat called *blubber*, which can be up to a foot and a half thick (about 46 cm). Blubber keeps whales warm in freezing temperatures and, since blubber is lighter than water, it allows whales to float better.

Whale pectoral flippers help stabilize a whale's body in the water. By slapping a flipper on the surface of the water, whales can communicate with each other.

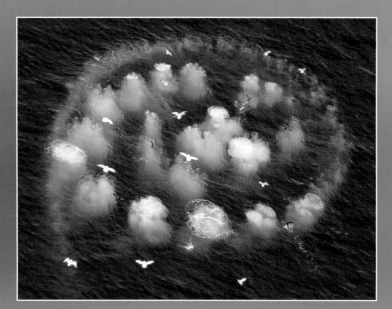

▲ BUBBLE NETTING

Some humpback whales cast their own nets when fishing—only their nets are made of bubbles! A group of whales dives deep below a school of fish. One whale blows bubbles while swimming in a circle. The bubbles rise up and form a cylinder of bubbles that the fish will not swim through. Then the other whales, with their mouths open wide, lunge to the surface through the middle of the cylinder. They get a huge mouthful of water and fish.

A whale's ears are very sensitive. Whales can hear other whale calls across hundreds of miles of ocean.

▼ WHALE PODS

Whales are social creatures that live in groups called *pods*. Some group members are old and others are young calves. A whale in a pod may live close to its parents for several years. Groups are especially effective when working together to round up fish with bubble nets, and for protecting the calves from predators. If predators threaten a sperm whale calf, the adults form a ring around the calf, facing outward. To get to the calf, the predator would probably think twice before attempting to get past the massive adults! Whales also produce amazing sounds to communicate with each other in breeding grounds, to hunt schools of fish, and to locate geographic features in the ocean as they migrate. Scientists have shown that whales can hear their calls through hundreds of miles of ocean.

The tail, or fluke, of a whale is very strong. It is used to propel the animal forward. Flukes also allow scientists to identify each animal by its unique tail markings.

SPOUTING ▶

Because they are mammals, whales need to breathe air. They must come to the surface. Whales breathe through blowholes located near the top of their head. Blue whales can hold air in their lungs for more than 30 minutes, and sperm whales can do it for up to 1.5 hours! When the whale exhales or spouts, the moist warm air from its lungs is released into the outside air, forming a cloud. Some spouts can reach a height of 33 feet (10 m).

did you know? BOWHEAD WHALES CAN LIVE CLOSE TO 200 YEARS!

GORILLAS

Gorillas are members of a group of primates known as the great apes. This group includes gorillas, chimpanzees, orangutans, and humans. Great apes are distinguished from monkeys by their larger size, upright posture, and lack of tails. Primates shared a common ancestor that lived more than 65 million years ago, but the gorilla line began only about 7 million years ago. Today, scientists recognize the several subspecies of gorilla, all living in equatorial Africa: the western lowland gorilla, the eastern lowland gorilla, and the mountain gorilla. Like all primates, gorillas have highly developed brains and a great capacity for learning. Researchers have taught gorillas sign language, which the gorillas have used to identify objects, numbers, words, and people. Researchers have also documented gorillas using sticks and stumps as tools in the wild. And, thanks to the work of dedicated scientists such as Dian Fossey, we also know a great deal about gorilla communication. These amazing animals express themselves using complicated vocalizations and gestures, including hoots, whines, chest beating, lip puckering, and smiling.

▲ SNOWFLAKE, THE WHITE GORILLA

Snowflake, a beautiful and rare white gorilla, was the first ape with a documented case of albinism. Albinism, which can also affect humans, is a genetic mutation that prevents the production of melanin—the pigment that colors our eyes, hair, and skin.

Older male eastern lowland gorillas are called "silverbacks" because their black hair turns silver below their shoulders.

Gorilla nose prints are like human fingerprints! Scientists can use them to identify individual gorillas.

Gorillas have five toes, including an opposable big toe that helps them grab branches and objects with their feet.

GORILLA REPRODUCTION ►

Gorillas have large, round bellies, making it difficult for scientists to determine whether a female is pregnant. Female gorillas give birth to one 4.5-pound (about 2 kg) infant gorilla after an 8½-month pregnancy. After about 4 years, she will give birth again. Young gorillas grow twice as fast as human babies and become independent when they are about 3½ years old!

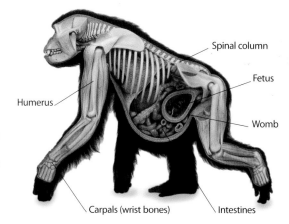

Spinal column

Fetus

Humerus

Womb

Carpals (wrist bones)

Intestines

PARENTAL CARE ►

Gorilla mothers invest a lot of energy in raising and protecting their young. For its first 5 months, a young gorilla never leaves its mother's side. The mother is responsible for feeding, grooming, and nesting with her offspring. The father teaches the young gorilla to interact and play with the rest of the troop.

Gorillas have binocular, color vision that lets them focus both eyes on an object and judge distance.

At about 6 or 7 months of age, young gorillas ride piggyback. They hang on by clutching their mother's long hair.

Unlike monkeys, apes do not have tails.

When gorillas walk, they use their arms (which are longer than their legs) for support, pressing their third and fourth knuckles against the ground. This is why they are known as "knuckle walkers."

◄ WESTERN LOWLAND GORILLAS

These gorillas live in the rain forests of central Africa and, like all gorillas, are highly endangered as a result of habitat destruction and poaching. Western lowland gorillas are distinguished by a wide, cone-shaped head, small ears, and short hair. They are shy vegetarians who use their strong jaws to eat more than 200 different types of plants. Male gorillas can eat up to 45 pounds (about 20.4 kg) of food per day!

FROZEN ZOO

Polar bears, seals, Arctic foxes—perhaps you can imagine finding these animals at a zoo somewhere in the frozen Arctic Circle. A *frozen zoo*, however, is a very different kind of zoo. You won't find animals there—but you will find their sperm, eggs, embryos, blood, cell cultures, and tissues. Samples like these contain the animals' genetic material. An individual animal's genetic material, or DNA, is what determines the animal's size, shape, color, and other physical characteristics. Genetic material can also support an entire species' ability to survive. Scientists have been collecting and freezing the genetic material of endangered animals for more than 30 years. Why? Conservation. Genetic material helps scientists protect and strengthen endangered animal populations. With DNA, scientists can assist with breeding the animals to increase their number of offspring and improve an entire species' chances of avoiding extinction.

know? did you
IN 1980, ONLY 19 CALIFORNIA CONDORS REMAINED IN THE WILD. WITH CONSERVATION EFFORTS AND HELP FROM SAN DIEGO'S FROZEN ZOO, THERE ARE NOW MORE THAN 300 BIRDS, 135 OF THEM LIVING IN THE WILD.

SAN DIEGO'S FROZEN ZOO ▼
Samples of genetic material from more than 8,400 individual animals from 800 species and subspecies "live" at the Frozen Zoo at San Diego Zoo's Institute for Conservation Research. They include animals like the Gobi bear, secretary bird, and African clawed frog. Scientists at the Frozen Zoo track evolutionary trends and preserve genetic variation, or differences among individuals within a species. Species with diverse gene pools have a greater chance of survival.

You might see a Bengal tiger cub in a zoo. Only between 3,000 and 4,700 of these endangered cats remain in the wild.

◄ THE GIANT PANDA CHALLENGE

About 1,600 giant pandas are all that remain in the wild, but more than 160 live in captivity. Pandas are extremely slow breeders. Females can bear young only about once every 2 years. To help find a solution to the problem, Chinese scientists used DNA from a panda's cells to create a clone, or exact copy of the panda, in the form of an embryo, or fertilized egg. Their plan is to place the cloned embryo inside a female that could carry and give birth to the panda.

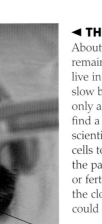

Giant panda cubs whose mothers are unable to care for them have been successfully raised by humans.

◄ YELLOW SEA HORSE

The yellow sea horse population is declining. The species is used in traditional Chinese medicine and sold for use in aquariums. Its DNA, along with the DNA or tissue samples of about 1,000 species, has been preserved by the Frozen Ark Project at the University of Nottingham in England.

Ocean pollution and habitat destruction also contribute to the decreasing number of yellow sea horses.

CRYOSPRESERVATION ►

Frozen zoos use cryopreservation, the process of freezing and preserving genetic material. Scientists fill special holding tanks with liquid nitrogen that keeps the temperature of the samples at –320°F (–196°C). Frozen genetic material can be stored indefinitely, perhaps for thousands of years. If a species nears extinction, scientists can thaw the samples and use them to help animals produce young.

113

EARTH SCIENCES

As far as we know, Earth is the only place in the universe capable of supporting life. This is due to unique features including an optimum distance from the sun, vast quantities of water, and a breathable and protective atmosphere. But our planet undergoes constant change due to natural or human-induced phenomena. Tectonic plate movement continually reshapes the surface by creating new continents, mountains, earthquakes, and volcanoes. Wild and unpredictable weather, floods, droughts, rock erosion, and pollution can have devastating effects. Studying the Earth reveals its past evolution, and helps us to predict how changes, like global warming, might affect its ability to sustain life in the future.

GEOLOGIC TIME

A volcano's flow of lava—melted rock from deep underground—gives us an idea of what Earth was like when it first formed.

A million years may seem like a long time to us, but it is only a tiny fraction of Earth's age. Scientists have calculated Earth to be 4.6 billion years old! Scientists have divided this time into units that we can use to describe the different parts of this immense span of time. To understand the length of Earth's history, imagine how it would look if it were drawn on the face of a clock. Say you mark the beginning of the geologic record of Earth—4.6 billion years ago—at 12:00 on the clock. A full circle of the clock, 12 hours, represents geologic time from the beginning of Earth to the present. In this model, each hour represents about 383 million years. You can use this model to picture the different lengths of time that spanned important events in Earth's history.

PRECAMBRIAN: 4.6 BILLION TO 542 MILLION YEARS AGO

Precambrian time is the longest era of the geologic record—more than 10 and one-half hours on the clock model. At the beginning, the planet's surface was mostly melted rock. It eventually cooled and formed a solid crust covered in part by oceans of liquid water. After more than 600 million years—between 1 and 2 hours on the clock—the first living things formed from chemical building blocks in these early oceans. Simple, single-celled bacteria used energy from sunlight and began producing oxygen. This oxygen helped create an atmosphere that would later support more complex life. It was nearly the end of Precambrian time before the first multicelled organisms appeared.

PALEOZOIC: 542 MILLION TO 251 MILLION YEARS AGO

In our clock model, the Paleozoic era took nearly an hour, extending to about 20 minutes after the 11 on the clock. It began with the Cambrian explosion, named for the diversity, or variety, of living things that appeared in that time. The continents came together to form giant land masses that later broke apart. Plants began to grow on land and many kinds of animals evolved. By the end of the Paleozoic, there were forests, amphibians, and reptiles that breathed with lungs. It ended with the largest mass extinction ever: the Permian extinction. About 70 percent of land animals and 90 percent of all ocean species were wiped out.

The oldest known trees had branches and leaves like those of these modern tree ferns.

did you know?
ALMOST THREE-QUARTERS OF THE PLANET'S 4.6 BILLION-YEAR HISTORY TOOK PLACE BEFORE THE SIMPLEST MULTICELLULAR LIFE FORMS EVEN CAME INTO EXISTENCE.

Modern humans first appeared about 190,000 years ago— less than 2 seconds in the clock model of geologic time!

MESOZOIC: 251 MILLION TO 66 MILLION YEARS AGO

The Mesozoic era lasted for about half an hour on the clock. When you think of extinct animals, you probably think of the dinosaurs. The dinosaurs ruled during this era. However, other living things evolved too. Mammals appeared about 200 million years ago. Birds began to fly about 50 million years after that. Plants also changed. Ferns, then cone-bearing trees, and later flowering plants and trees became the dominant plants on land. The Mesozoic ended with another mass extinction that killed off the dinosaurs. Their absence created an opportunity for modern animals to evolve and flourish.

CENOZOIC: 66 MILLION YEARS AGO TO THE PRESENT

If you imagine that Earth's history took place in a 12-hour period, the entire Cenozoic era happened during only the last 10 minutes. During this era, small mammals and reptiles thrived. As the continents moved closer to where they are today and the climate went through many changes, plants and animals began to evolve with the shifting conditions. Whales evolved from land mammals that moved to the oceans. Grasslands formed where forests retreated. Large grazing animals appeared. Only in the last 30 seconds of the clock model did the ancestors of humans begin to walk on two legs and use tools.

FOSSILS

Fossils are like Earth's history book. They can show what lived where and when, and who ate what—or whom. Fossils have been discovered in some very unlikely places. Saber-toothed cats, mammoths, ground sloths, and prehistoric American lion fossils have been found in the middle of the city of Los Angeles! These fossils came from the site of La Brea Tar Pits, the largest and most diverse collection of plants and animals from the Ice Age. Fossils of more than 2,000 individual saber-toothed cats have been recovered from this site. But these fossils are very young—only 10,000 to 40,000 years old. Some of the oldest known fossils, which are of bacteria, are nearly 3.5 billion years old.

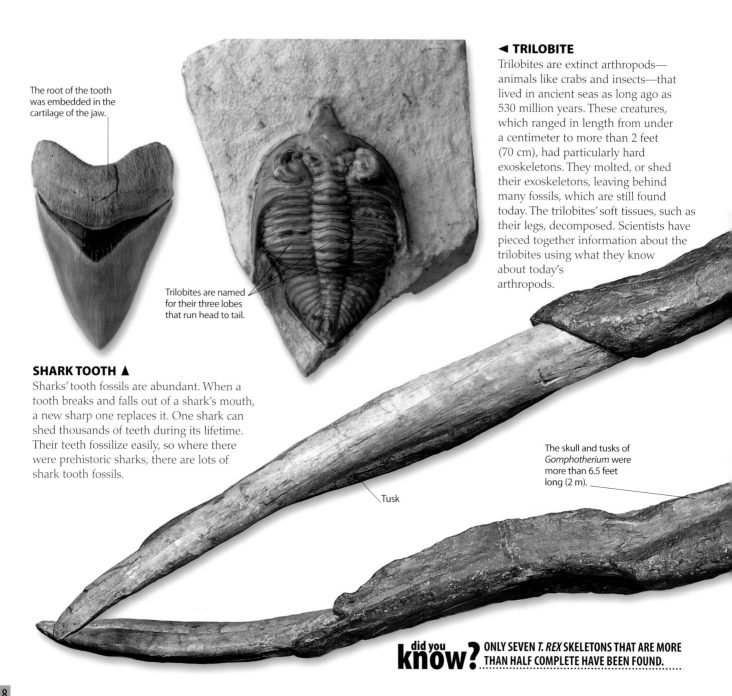

The root of the tooth was embedded in the cartilage of the jaw.

Trilobites are named for their three lobes that run head to tail.

◄ TRILOBITE

Trilobites are extinct arthropods—animals like crabs and insects—that lived in ancient seas as long ago as 530 million years. These creatures, which ranged in length from under a centimeter to more than 2 feet (70 cm), had particularly hard exoskeletons. They molted, or shed their exoskeletons, leaving behind many fossils, which are still found today. The trilobites' soft tissues, such as their legs, decomposed. Scientists have pieced together information about the trilobites using what they know about today's arthropods.

SHARK TOOTH ▲

Sharks' tooth fossils are abundant. When a tooth breaks and falls out of a shark's mouth, a new sharp one replaces it. One shark can shed thousands of teeth during its lifetime. Their teeth fossilize easily, so where there were prehistoric sharks, there are lots of shark tooth fossils.

Tusk

The skull and tusks of *Gomphotherium* were more than 6.5 feet long (2 m).

did you know? ONLY SEVEN *T. REX* SKELETONS THAT ARE MORE THAN HALF COMPLETE HAVE BEEN FOUND.

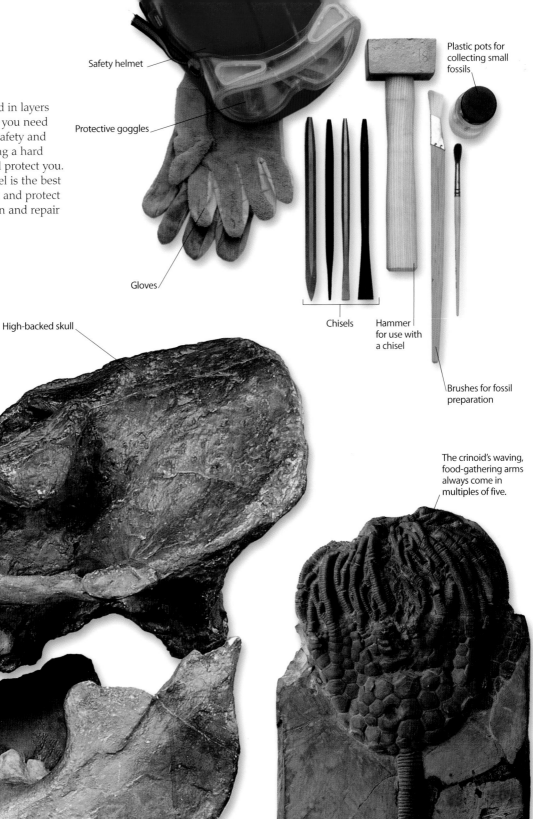

FOSSIL HUNTING ►

Fossils are most often buried in layers of rock. To get to the fossils, you need the proper tools—for your safety and for the fossil's safety. Wearing a hard hat, goggles, and gloves will protect you. Using a hammer and a chisel is the best way to get through the rock and protect the fossil. Brushes help clean and repair the fossil.

Safety helmet

Protective goggles

Gloves

Plastic pots for collecting small fossils

Chisels

Hammer for use with a chisel

Brushes for fossil preparation

High-backed skull

Narial opening (nostril)

Cheek tooth

The crinoid's waving, food-gathering arms always come in multiples of five.

GOMPHOTHERIUM ▲

This skull from the *Gomphotherium* genus is about 20 million years old. *Gomphotheres* were ancestors of extinct mastodons and modern elephants. They had both upper and lower tusks, and most likely lived in lakes and swamps where they used their lower tusks to dig up the vegetation. *Gomphotherium* remains have been found in many parts of the world, including Germany, Kenya, and even in the middle of the United States, in Kansas.

FOSSIL CRINOID ▲

Although it looks like a plant, a crinoid is actually an animal called an *echinoderm*. It is a relative of sea urchins and sea stars. During the Paleozoic era, crinoids blanketed the sea floor. Experts have used their fossil records to identify hundreds of different species of crinoids, some of which still exist.

DINOSAURS

Who says that dinosaurs are ancient history? Even now, scientists are discovering new things about these Mesozoic beasts! In 1995, a fossil hunter in Argentina discovered the skeleton of a new species of dinosaur. Scientists measured its bones and found that this giant meat-eating creature was bigger than *T. rex*! Maybe that's why they named it *Giganotosaurus*. Along with new discoveries, there are also new arguments brewing among scientists. Paleontologists, the scientists who study dinosaurs, are arguing about whether dinosaurs were warmblooded or coldblooded. Recent discoveries show that some dinosaurs were quick and active, and not at all like the slow, lumbering, coldblooded reptiles that dinosaurs were once thought to be. Now, many scientists are also saying that dinosaurs are not really extinct. Most paleontologists believe that modern birds, such as ostriches, are actually related to some of the dinosaurs who lived 100 to 200 million years ago.

Up in the nose of many dinosaurs were big, fleshy nasal passages. These helped the animals gain or lose heat when they breathed air.

These sharp, sawlike teeth were perfect for tearing meat.

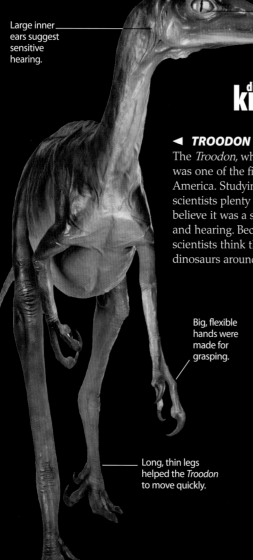

Large inner ears suggest sensitive hearing.

Large eyes suggest well-developed vision.

did you know? PLANT-EATING DINOSAURS COULD GROW OVER 100 FEET (30 M) LONG AND WEIGH AS MUCH AS 33 CARS! ....

◀ TROODON

The *Troodon*, whose name means "wounding tooth," was one of the first dinosaurs discovered in North America. Studying the *Troodon's* skeleton has given scientists plenty of clues about what it was like. They believe it was a small, quick predator with keen vision and hearing. Because of the large size of its braincase, scientists think the *Troodon* was one of the smartest dinosaurs around.

Big, flexible hands were made for grasping.

Some pterosaurs had flaps on their tails to keep them stable during flight.

Long, thin legs helped the *Troodon* to move quickly.

PTEROSAUR ▶

It's not hard to figure out why the pterosaur's name means "wing lizard." However, it may be surprising to know that these flying creatures were not dinosaurs, even though they lived during the same time. Dinosaurs were land animals who never flew until some of their descendants used feathers to fly. Pterosaurs flew using wings with no feathers. They were relatives of the dinosaurs who ranged in size from tiny birds to airplanes. Their bones were hollow, though, so these lightweight winged lizards could soar easily through the skies.

▼ GIGANOTOSAURUS

The *Giganotosaurus* is part of a group of meat-eating dinosaurs called *theropods,* whose name means "beast feet." Scientists say that the *Giganotosaurus* is probably the largest theropod in the world. Its body measures about 41 feet (12.5 m) long and it is estimated to have weighed between six and eight tons!

Dinosaur fossils sometimes show the shape and size of horns and armor.

This is what scientists believe the *Giganotosaurus's* tongue looked like.

The neck is protected by thick, dry, scaly skin.

A hairlike body covering suggests that some pterosaurs may have been warmblooded.

Catching fish from the sea was easy with this scooplike beak.

The wing was supported along its edge by a very long fourth finger from the hand.

EXTINCTION

Imagine an asteroid the size of Manhattan hurtling toward Earth, its edges in flames as it burns through the atmosphere. If you think such an event happens only in science fiction, think again. Scientists have found evidence that large asteroids have hit Earth in the past. Many hypothesize that this kind of event caused the mass extinction of the dinosaurs. Extinction happens when an entire species dies out. A mass extinction occurs when hundreds of different species become extinct in a single event. The largest mass extinction took place about 250 million years ago, just before the first dinosaurs came into existence. This event, called the *Permian–Triassic extinction,* killed off more than 90 percent of all sea life and 70 percent of land animals.

▼ CAUSES OF MASS EXTINCTION

At least five mass extinctions have taken place in the last 540 million years. These may have been caused by asteroids or comets colliding with Earth, increased volcanic activity, ice ages, or changes in sea level. Species that survive have a new chance at life. After the Permian-Triassic extinction, new species took the places of those that disappeared. Dinosaurs, such as *Corythosaurus* and *Ceratops*, dominated the era that followed.

Plant-eating *Corythosaurus* had a toothless bill like a duck and was 30 feet (about 9 m) long.

Members of the *Ceratops* genus had beaklike mouths and bony frills on their heads. Some had horns. They probably ate plants.

AN ENDANGERED SPECIES IS DEFINED AS ONE THAT IS IN DANGER OF EXTINCTION THROUGHOUT ALL OR A SIGNIFICANT PORTION OF ITS RANGE.

◄ SURVIVING AN ASTEROID COLLISION

Species that survive an asteroid collision confront other problems. Dust sent up when the asteroid hits Earth could darken the sky for weeks. The dust can strain breathing, harm plants that depend on sunlight, and kill single-celled organisms that are sensitive to even subtle chemical changes in their surroundings. The extinction of some species can cause the death of others that depend on them for food and oxygen.

Tyrannosaurus bones found in the Rocky Mountain region of North America show bite marks made by *T. rex* teeth, leading scientists to believe they fought one another or were cannibals.

Scientists believe that *T. Rex* was wiped out during the Cretaceous-Tertiary mass extinction, which may have been caused by the Yucatán Peninsula asteroid.

▲ WAS *T. REX* IN YOUR BACK YARD?

Not all extinction hypotheses can be proved, leaving scientists with incomplete answers. Fossils provide the strongest evidence of extinction. Scientists have also observed a pattern in fossils that suggests mass extinctions occur every 26–30 million years. Some believe the pattern relates to regular travel paths of asteroids and other celestial objects.

GIANT MAMMALS

The last 65 million years are known as the Cenozoic era. This era is also known as the Age of Mammals, because mammals thrived during this time. Many mammals grew to incredible sizes during the Cenozoic era. These giant mammals included beavers almost the size of modern day black bears, and cave lions that weighed around 800 pounds (360 kg)! Most giant mammals became extinct around 10,000 years ago for reasons that are still being debated. Some scientists believe early humans hunted the giant mammals to extinction. Other scientists blame the last ice age, which began 70,000 years ago when glaciers—large, moving sheets of ice—spread across much of Earth limiting greatly the resources these animals needed to survive.

WOOLLY MAMMOTH ▶

The woolly mammoth—a symbol of the last ice age—is probably the most famous Cenozoic mammal. Scientists know a lot about them from preserved mammoth carcasses found in Siberia and from European cave drawings. Woolly mammoths dominated the cold, northern regions of Europe, Asia, and North America between 350,000 and 10,000 years ago. They were about 11 feet (about 3 m) tall and may have eaten about 300 pounds (136 kg) of vegetation a day!

A two-layered coat kept mammoths toasty warm. Long, shaggy guard hairs covered a layer of dense underfur.

Woolly mammoths had curving ivory tusks that could reach lengths of 10 feet (about 3 m).

A mammoth stands 11 feet tall; this man is 5'9" tall.

did you know?
THE NOW EXTINCT HORNLESS RHINOCEROS *INDRICOTHERIUM* WAS THE LARGEST LAND MAMMAL EVER! IT WAS 26 FEET (ALMOST 8 M) LONG, 18 FEET (ABOUT 5.5 M) TALL, AND WEIGHED ABOUT 20 TONS.

ARSINOITHERIUM ▶

This rhinoceros-like African herbivore lived in swamps around 30 million years ago. A male arsinoitherium had two huge horns sticking out from the front of its head. These horns were hollow and scientists believe they may have been used to produce mating calls and to duel with other males. These massive mammals stood about 6 feet (1.8 m) at the shoulder.

The giant sloth's head was small compared with its body.

Arsinoitherium stands about 6 feet tall; this man is 5'9" tall.

◀ GIANT SLOTH

The king of the Cenozoic ground sloths was the South American mammal *Megatherium*. As big as a modern elephant, the long-haired *Megatherium* resembled an overgrown guinea pig with a long tail. Many scientists think that *Megatherium* could stand on its hind legs and dine on leaves and twigs that it pulled off the very top branches of trees with its long, sharp claws.

A giant sloth stands 20 feet tall; this man is 5'9" tall.

Megatherium may have used its strong tail for balance and support when it stood on its back legs to reach leaves.

DATING ROCKS

A tree has rings, but how do you tell the age of a rock? Rocks have built-in clocks. Some elements in rocks go through a process called *radioactive decay.* The atoms of these elements emit particles from their nuclei. Over millions of years, this decay causes one type of atom to become another type. Take the element uranium, for example, a radioactive metal found in many rocks. Some forms of uranium decay into the element lead. If geologists count the atoms of lead in a rock sample, and compare it with the number of uranium atoms, they can tell how old the rock is. How? In the same way that we know how long it takes a log to burn, geologists know how long it takes uranium to decay into lead. So, the more lead in the rock, the older it is, because it has spent more time decaying.

did you know?

THE OLDEST KNOWN ROCK FORMATIONS ON EARTH CAN BE FOUND IN GREENLAND. THEY ARE THOUGHT TO BE 3.8 BILLION YEARS OLD.

DATING CAVE PAINTINGS ▼

Scientists can also date matter that was once living, such as bones and fabric. In the Pech Merle cave in France, the prehistoric artists used tiny amounts of charcoal in their black paint. Scientists measured the small amount of the radioactive form of carbon, carbon-14, that was in the charcoal to estimate that the paintings were created about 25,000 B.C. Scientists use different radioactive elements depending on what elements make up the item being dated. For instance, potassium-40 is used to date moon rocks, because they contain trace amounts of this radioactive element.

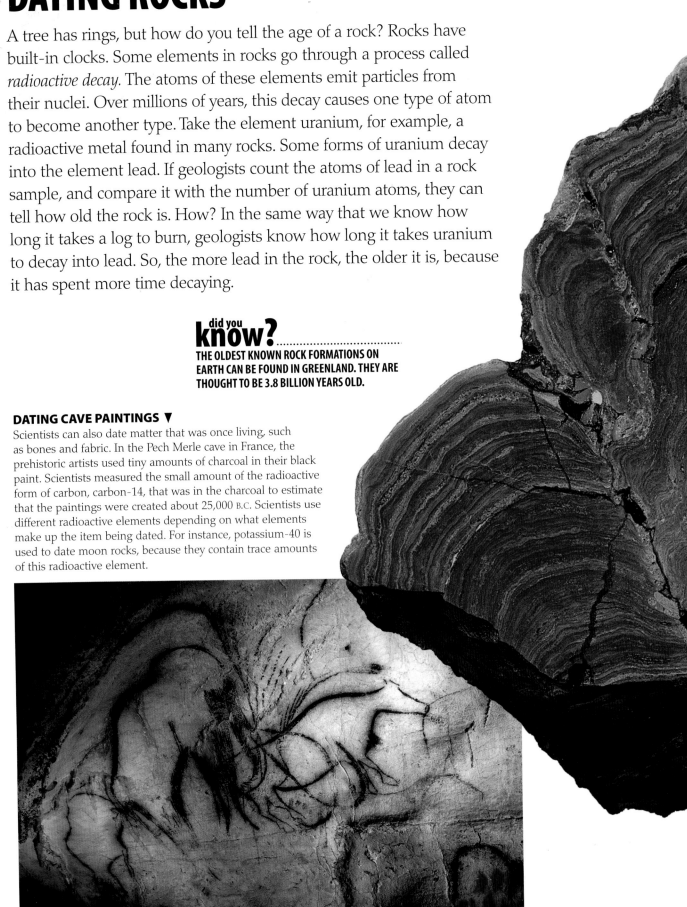

RESEARCH TOOLS ▲

Radioactive elements have what is called a *half-life*: the length of time it takes for half the atoms in element A to decay into element B. Using a tool called a *mass spectrometer*, a scientist finds that half of the atoms in a sample are A and half are B. If A has a half-life of 100 years, and half its atoms have decayed into B, the rock sample must be 100 years old. The elements used to date rocks have incredibly long half-lives. Uranium-238, for example, has a half-life of about 4.5 billion years!

Stromatolites are rocklike structures made of layers of primitive microorganisms, such as algae and bacteria, pressed together with layers of mineral deposits. Fossil stromatolites represent some of the first forms of life on Earth and have been found in rocks dated at 3.5 billion years old.

Some moon rocks collected by Apollo astronauts ranged from 3 to 4.6 billion years old.

A pair of fossilized tree cones dates from between 65 and 145 million years ago.

LAYERS UPON LAYERS ▲

Scientists often use more than one technique to date rocks. Rock layers provide a way of learning what's called the *relative age* of rocks and fossils. Older rocks are usually in the bottom layers, while younger rocks are on top. This helps scientists infer the relative age of fossils that are found in different layers of rock.

COAL

Did you know that you can use dead plants to turn on your lights? Well, kind of. Some electricity in the United States is produced by coal, which, at its basic level, is made from dead plants—really dead ones—millions of years old. When layer upon layer of plant remains are compressed under rocks and dirt for millions of years, the result is coal—a brownish-black sedimentary rock that burns. Coal is made up of carbons and hydrocarbons, which are quite combustible. The energy that was trapped in the plants underneath all the rocks and soil is released when the coal is burned. Power plants burn coal to make steam, and the steam turns turbines that produce electricity. So dead plants = working light bulbs.

SURFACE MINING ▼

When coal is buried less than 200 feet (about 61 m) underground, it can be surface mined. Most of the coal in the United States is mined this way. Monstrous machines remove the topsoil and expose the coal beds beneath. After the coal is mined, the topsoil is replaced. This coal mine is in Wyoming—where about 40 percent of the nation's coal is mined.

did you know? U.S. COAL DEPOSITS CONTAIN MORE ENERGY THAN ALL THE OIL IN THE WORLD.

Contour mining, a type of surface mining, is used in mountainous areas. The paths follow coal beds along the hills.

HOW COAL FORMED Coal is a fossil fuel, made from the remains of ancient animals and plants. Coal is also a nonrenewable resource. Once it is used up, it cannot be replenished.

From 100 to 300 million years ago, giant plants containing lots of energy lived in swampy forests.

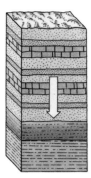

The giant plants died and were buried. As they decayed, they formed a material called *peat*.

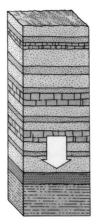

Over millions of years, the pressure exerted by accumulated sediments and heat compressed the peat into different types of coal such as lignite, anthracite, subbituminous, and bituminous coal.

The coal bed is divided up and mined in layers.

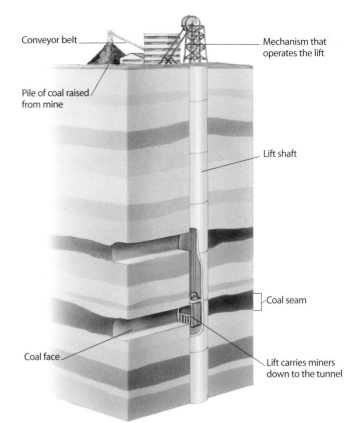

Conveyor belt

Pile of coal raised from mine

Mechanism that operates the lift

Lift shaft

Coal seam

Coal face

Lift carries miners down to the tunnel

DEEP MINING ▲

Coal is sometimes buried more than 200 feet (61 m) underground. Brave miners take elevators down deep mine shafts. They cut the coal from the coal face with huge machinery and send it up on long conveyor belts.

ICE AGE

Would you believe someone who says we are living in an ice age? You should. Ice ages can last many thousands of years. An ice age consists of both cold periods (glacials), marked by the widespread advance of glaciers, followed by warm periods (interglacials), when glaciers retreat. We are living in an interglacial period. The last glacial period ended about 12,000 years ago during the Pleistocene epoch. In that glacial period, mountain ice caps and sea ice grew. Thick ice sheets extended from the north polar region into Greenland, Russia, and northern Eurasia, and covered Canada and parts of the United States. Ancient ice still exists in Greenland, Antarctica, and some mountain ranges. Scientists study the ice and the grounds where glaciers once were to learn how ice ages affect Earth's climate and geology. Fossilized animals and plant matter, sediment and ice cores, and rock studies all provide evidence of prehistoric life and climate change.

READING AN ICE AGE

Scientists can determine Pleistocene glacial activity by studying rock layers, rock surfaces marked by glacial movement, and boulders carried great distances by advancing ice sheets. There are many fossils from the Pleistocene epoch. Because they are well preserved, scientists can tell exactly when the organisms lived. Fossils of plant pollen, single-celled organisms, and animal remains provide significant data about the effects of climate change on plant and animal life.

Fossilized bone remains

THE WOOLLY MAMMOTH ▲

The woolly mammoth is the best-known fossilized vertebrate animal from the Pleistocene epoch. Carcasses, skeletons, and other fossils have been found in Eurasia and North America. Scientists debate why the species became extinct. Climate change, overhunting, and disease are three strong theories.

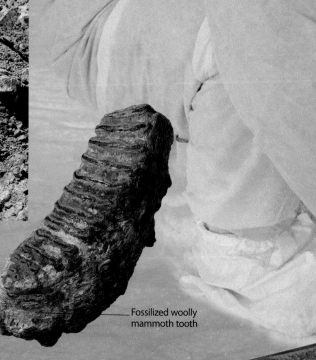

Fossilized woolly mammoth tooth

◄ EARTH'S OWN CLIMATE RECORDS

Ancient rocks, dating to about 2.3 billion years ago, contain evidence that a rise in atmospheric oxygen led to the earliest ice age. Sediment tells scientists where glaciers were located. Fossils of pine cones, stems, leaves, and pollen indicate how the movement of glaciers, as well as the climate change it caused, affected where different northern trees grew. Tree rings of fossilized ice age wood show how much growth took place.

Tree ring widths reflect changes in rainfall and temperature from one growing season to the next.

Face masks protect the ice core sample from germs and heat carried by the scientist's breath.

GLACIAL ICE CORE STUDY ►

Scientists study glacial ice cores to determine how the composition of the atmosphere has changed over time. They extract gases from tiny air bubbles trapped in the core and measure how much carbon dioxide, methane, and other greenhouse gases they contain. They also infer historic temperatures by studying water molecules released when the core melts. Scientists use both sets of data to reconstruct Earth's climate change record.

Ice core sample

Ice core drill bit

did you know?.............
TO STUDY HUNDREDS OF THOUSANDS OF YEARS OF CLIMATE HISTORY, SCIENTISTS HAVE DRILLED ICE CORE SAMPLES DEEPER THAN 2 MILES (3.3 KM)!

GLACIERS

Earth's poles are locked in ice, but for how long? Glaciers—large masses of ice that grow and move over time—have advanced and retreated throughout Earth's history. As they do, they leave behind telltale landforms as signs of their movement. Deposits of rock left by glaciers in South Africa and Australia, which are 290 million years old, provide evidence that these areas that are now separate were once joined. At that time, much of the world was covered in ice sheets. During the reign of the dinosaurs, 145 million years ago, the world was warmer. There were no ice sheets on Earth, not even at the poles! Sea levels were higher and a shallow inland sea covered what is now the Great Plains of the United States. Cooler temperatures eventually returned, glaciers formed again, and sea levels fell. Today, three fourths of Earth's fresh water is frozen in ice caps and glaciers. As Earth's average temperature rises, glacier ice melts faster than it accumulates. Will shallow inland seas return some day?

CAVES AND CALVING ▼

Glacier ice can be unstable where a glacier meets the sea. The vertical cracks, called *crevasses*, in this glacier form at areas of weakness. Meltwater, running down into a crevasse, has widened one crack far below into an ice cave. Often, huge chunks of the glacier break free and fall into the sea below becoming icebergs. This process, called *calving*, increases as global temperatures rise.

did you know?

THE GLACIER ICE IN ANTARCTICA IS MORE THAN 2.5 MILES (4 KM) THICK!

Accumulation zone, where snow builds up from winter to winter

Glacier's end, called its *terminus*

◄ A GLACIER FORMS

Glaciers form when snow falls and stays frozen from one winter until the next. Each year, more snow accumulates, making a larger snowfield that reflects the summer sunlight. Reflected sunlight reduces the amount of melting before winter arrives again. Years of accumulation, along with melting and recrystallizing of snow, creates a mass of ice that advances downhill under its own weight—a glacier.

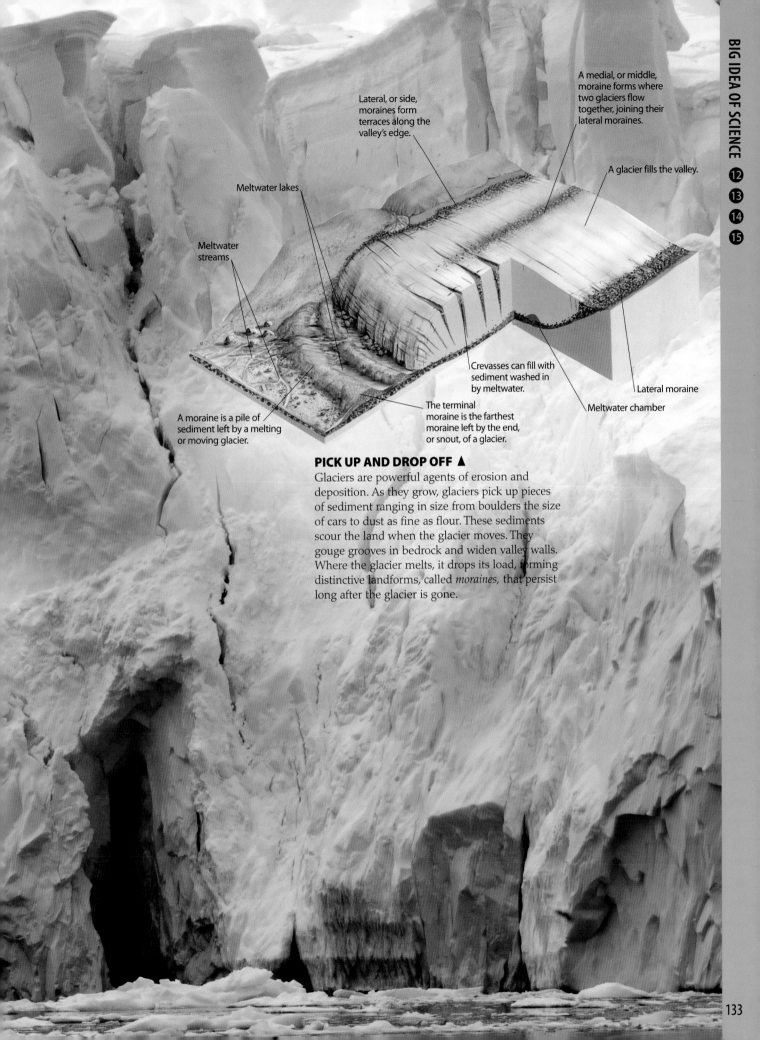

Lateral, or side, moraines form terraces along the valley's edge.

A medial, or middle, moraine forms where two glaciers flow together, joining their lateral moraines.

A glacier fills the valley.

Meltwater lakes

Meltwater streams

Crevasses can fill with sediment washed in by meltwater.

Lateral moraine

Meltwater chamber

A moraine is a pile of sediment left by a melting or moving glacier.

The terminal moraine is the farthest moraine left by the end, or snout, of a glacier.

PICK UP AND DROP OFF ▲

Glaciers are powerful agents of erosion and deposition. As they grow, glaciers pick up pieces of sediment ranging in size from boulders the size of cars to dust as fine as flour. These sediments scour the land when the glacier moves. They gouge grooves in bedrock and widen valley walls. Where the glacier melts, it drops its load, forming distinctive landforms, called *moraines,* that persist long after the glacier is gone.

WATER

Each of us is made up of roughly 60–70 percent water. In fact, all life on Earth owes its existence to water, a molecule that consists of two atoms of hydrogen bound to one side of an atom of oxygen. Water is a polar molecule: the oxygen end of the molecule has a slightly negative charge and the hydrogen end has a slightly positive charge. This causes water molecules to be attracted to each other. It also causes the particles of many substances to separate from each other in water, a process that is called *dissolving*. Because of water's ability to dissolve so many substances, water is essential for our body's health. Most nutrients in the body are dissolved in water and transported to cells. Wastes are dissolved in water and transported away from cells. Water helps in regulating body temperature. Red blood cells carrying oxygen are suspended in blood, the liquid portion of which is a solution of more than 90 percent water. In addition, water plays an important role in several chemical processes in the body. Water is needed to break down large molecules, such as proteins, into smaller molecules, like amino acids, that the body can use. Water is produced as a by-product when small molecules combine to form large molecules, such as starches and proteins. Water is also produced during cellular respiration, the process by which a cell breaks down glucose to get energy.

Fresh water is found in rivers, lakes, and streams. Water is also found below the Earth's surface.

Much of Earth's fresh water is frozen as ice and snow in the polar ice caps and glaciers.

Most of Earth's water is found in the ocean. Many salts, such as the salt you sprinkle on food, are dissolved in ocean water.

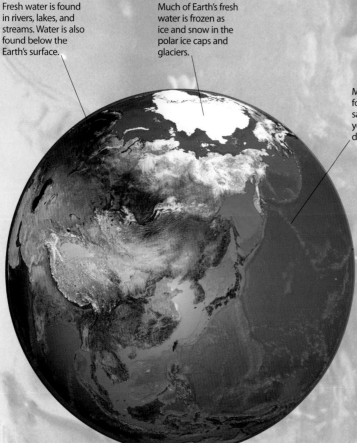

◄ HOW MUCH WATER?

About 5 hours into their trip to the moon, the astronauts on board Apollo 17 took several pictures of Earth. Earth appears blue because almost three fourths of its surface is water. Almost 97 percent of Earth's water is in the oceans. Most of the fresh water is frozen in the polar ice caps, glaciers, and snow. That leaves about 1 percent of Earth's water in rivers and lakes, swamps, and groundwater sources. There is also water in the air, some appearing as clouds, and in the cells of living things.

Water remains liquid over a wide range of temperatures. This seal lives in near-freezing waters.

Water flowing over a rocky cliff edge and pounding the rocks below can slowly dissolve minerals in the rock.

ICE ▲

An unusual property of water is that its solid state—ice—floats on top of its liquid state—water, since ice is less dense than water. When water freezes, ice forms from the surface downward. Unlike fish, aquatic mammals such as this Weddell seal must come to the surface to breathe air. This seal has found a breathing hole in the ice.

did you know?
....

IF EVENLY SPREAD OVER ALL LAND SURFACES, SALT FROM ALL THE OCEANS WOULD FORM A LAYER MORE THAN 500 FEET (152 M) THICK.

WATERFALL ►

Water is the only natural substance that is found on Earth in all three states: solid, liquid, and gas. Water changes from one state to another in what is called the *hydrologic cycle*. For example, liquid water in lakes, rivers, and oceans evaporates to form water vapor. In the atmosphere, this gas condenses and returns to its liquid state. This constant cycle redistributes water on Earth's surface. Water then shapes the land by dissolving rock and depositing sediments.

OCEAN CURRENTS

The water in the ocean is in constant motion. It moves in huge continuous streams we call *currents*. When you see waves crashing against the shore, you are looking at the power of ocean currents. But how do they form? Ocean currents are of three basic types: surface currents, deep currents, and tidal currents. They are caused by wind, gravity, and water density, and are also affected by the position of the continents. Surface currents occur in the top 328 feet (100 m) of the ocean and are driven mainly by wind. They have a big impact on Earth's weather and climate. Deep currents sweep along the seafloor and are driven by water density and gravity. Tides, caused by the gravitational pull of the moon and sun, move water up and down. Although tidal currents affect smaller areas than other ocean currents do, they are important because they affect life, transport, and commerce along the coasts.

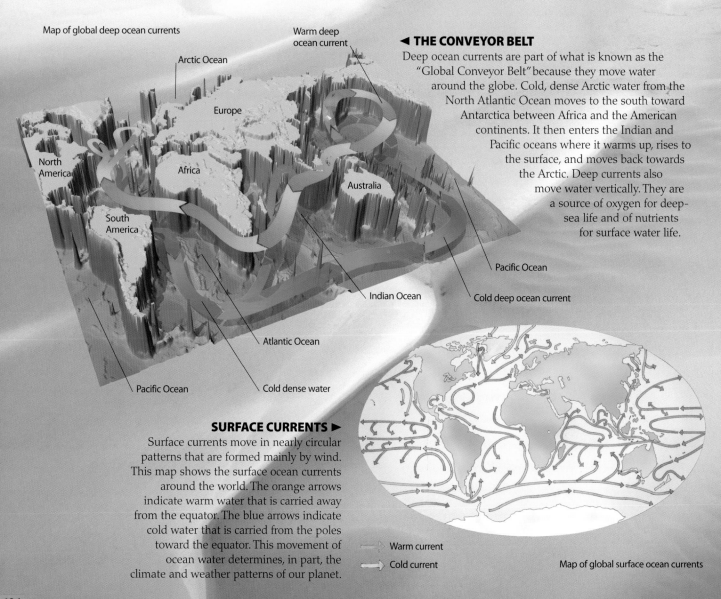

Map of global deep ocean currents

Warm deep ocean current

Arctic Ocean

Europe

North America

Africa

Australia

South America

Pacific Ocean

Indian Ocean

Pacific Ocean

Cold deep ocean current

Atlantic Ocean

Cold dense water

◄ THE CONVEYOR BELT

Deep ocean currents are part of what is known as the "Global Conveyor Belt" because they move water around the globe. Cold, dense Arctic water from the North Atlantic Ocean moves to the south toward Antarctica between Africa and the American continents. It then enters the Indian and Pacific oceans where it warms up, rises to the surface, and moves back towards the Arctic. Deep currents also move water vertically. They are a source of oxygen for deep-sea life and of nutrients for surface water life.

SURFACE CURRENTS ►

Surface currents move in nearly circular patterns that are formed mainly by wind. This map shows the surface ocean currents around the world. The orange arrows indicate warm water that is carried away from the equator. The blue arrows indicate cold water that is carried from the poles toward the equator. This movement of ocean water determines, in part, the climate and weather patterns of our planet.

Warm current

Cold current

Map of global surface ocean currents

California

Hawaiian Islands

EL NIÑO ▲

El Niño is the periodic irregular warming of water from the coasts of Ecuador and Peru to the central Pacific Ocean. During an El Niño period, the winds become weaker. This allows warm currents to flow from the west, heating the usually colder surface of the ocean. The yellow, orange, and red in this map of the Pacific Ocean indicate the warm waters of El Niño. El Niño is thought to have contributed to some torrential rainfalls in South America and the United States in the last couple of decades, as well as drought in Australia and record high temperatures in Europe.

Rip currents, which carry tidal waters back toward the sea, can be a hazard to swimmers.

TIDES

A tide is an alternating high and low point in sea level with reference to land. Tides are produced mainly by the gravitational pull of the moon, and to a lesser extent by that of the sun. Twice a day, water pushes onto the shore and then flows away in a predictable way. The current produced by a high tide is known as a *flood current*. The current produced by a low tide is called an *ebb current*. Tides can determine when ships can enter port and when or where fishers can expect a better catch. Tides are an essential part of the daily lives of coastal peoples.

Tidal currents and surface currents move sand around, creating dangerous sandbars and deep channels.

Mariners in small boats or large ships must be aware of the power of currents.

MID-OCEAN RIDGE

Some of the most dynamic parts of Earth's surface are also some of the least known. Mid-ocean ridges occur where Earth's tectonic plates are stretched and pulled apart. Oceanic plates move away from one another at amazingly slow rates of about 0.4 to 8 inches (1 to 20 cm) per year. The movement makes an opening down to the hot magma below. As the plates separate, this magma, or molten rock, bubbles up through the opening, creating mountain ranges. Almost all of these mountains are completely under water. They form a ridge system that winds around the planet in a chain that is more than 40,000 miles (about 65,000 km) long. On average, the top of the ridge is more than 1¼ miles (2,000 m) beneath the ocean's surface, deeper than the Grand Canyon. In some places, such as Iceland, the ridge extends above sea level, and appears as islands with volcanoes.

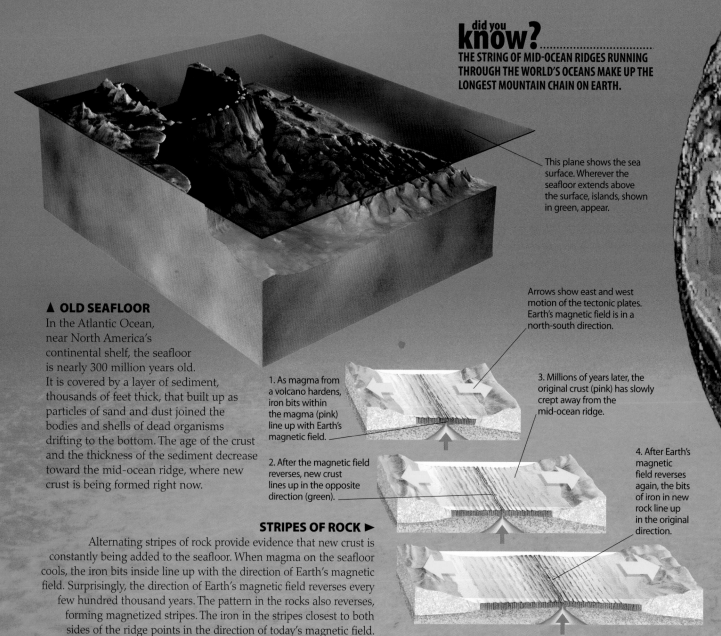

This plane shows the sea surface. Wherever the seafloor extends above the surface, islands, shown in green, appear.

Arrows show east and west motion of the tectonic plates. Earth's magnetic field is in a north-south direction.

1. As magma from a volcano hardens, iron bits within the magma (pink) line up with Earth's magnetic field.

2. After the magnetic field reverses, new crust lines up in the opposite direction (green).

3. Millions of years later, the original crust (pink) has slowly crept away from the mid-ocean ridge.

4. After Earth's magnetic field reverses again, the bits of iron in new rock line up in the original direction.

▲ OLD SEAFLOOR

In the Atlantic Ocean, near North America's continental shelf, the seafloor is nearly 300 million years old. It is covered by a layer of sediment, thousands of feet thick, that built up as particles of sand and dust joined the bodies and shells of dead organisms drifting to the bottom. The age of the crust and the thickness of the sediment decrease toward the mid-ocean ridge, where new crust is being formed right now.

STRIPES OF ROCK ▶

Alternating stripes of rock provide evidence that new crust is constantly being added to the seafloor. When magma on the seafloor cools, the iron bits inside line up with the direction of Earth's magnetic field. Surprisingly, the direction of Earth's magnetic field reverses every few hundred thousand years. The pattern in the rocks also reverses, forming magnetized stripes. The iron in the stripes closest to both sides of the ridge points in the direction of today's magnetic field. Stripes of older rock have been pushed farther from the ridge.

NEW SEAFLOOR

The Mid-Atlantic Ridge extends from the Arctic, through Iceland, and south nearly to Antarctica, dividing the Atlantic Ocean. Along this ridge, seafloor spreading occurs—new crust forms as magma oozes through weak places in the crust. As this crust moves east and west, it cools and slowly sinks so that the ocean on either side of the ridge is much deeper. You might think that the Mid-Atlantic Ridge is a place with no life. This is not the case, however. Hydrothermal vents provide hot water, rich in minerals from magma. An entire ecosystem lives near these places in the ridge where the minerals in the heated water provide energy for living things.

The depth of the ocean is shown by color—darker blue is deeper water. The Mid-Atlantic Ridge shows up as a light blue line splitting the seafloor.

Continental shelf

Iceland

Antarctica

DEEP SEA VENTS

Deep sea vents are like rocky, underwater chimneys. In fact, they are often called *smokers*, because they spew plumes of hot, mineral-rich fluid that look like chimney smoke. They occur thousands of feet below the ocean surface, around the places where tectonic plates—huge pieces of Earth's crust—move away from one another. These places are called *mid-ocean ridges*. The vents form when cold water sinks into cracks around these ridges. Magma, molten rock deep within Earth, heats the water, which then rises back up through the ocean floor. Even though the sun's rays cannot reach deep sea vents, hundreds of organisms live around them. How? Instead of converting sunlight into energy through photosynthesis, these deep sea vent communities get energy through chemicals coming out of the vents, in a process called *chemosynthesis*. Could this be how organisms first developed on Earth?

WHAT'S FOR DINNER? ▼

The image below compares photosynthesis and chemosynthesis. The chemosynthetic cycle begins with bacteria. Bacteria feed on chemicals released by the vents. Smaller animals prey on these bacteria and then larger animals feed on the smaller ones.

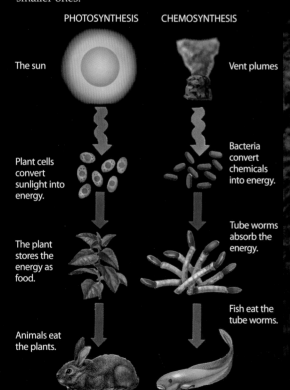

PHOTOSYNTHESIS   CHEMOSYNTHESIS

The sun   Vent plumes

Plant cells convert sunlight into energy.   Bacteria convert chemicals into energy.

The plant stores the energy as food.   Tube worms absorb the energy.

Animals eat the plants.   Fish eat the tube worms.

TOTALLY TUBULAR WORMS ►

Tube worms look like long tubes of red lipstick. They are huge animals—up to 8 feet (about 2.4 m) long, although most of their bodies are usually hidden in the tubes. They have a special relationship with bacteria. The bacteria live inside the worms, turning chemicals from the vents into nutrients for the worms. The worms in turn may become food for fish and crabs. Yum!

This is a black smoker. Smokers can also be white, clear, or gray, depending on the minerals that are dissolved in the fluid.

◄ MARINE LIFE AROUND A VENT

Deep sea vents form a unique ecosystem for the creatures that live there. As a matter of fact, it is one of the most extreme environments on Earth. The creatures that live in vent communities have adapted to live in total darkness. They endure freezing-cold water, scorching-hot vents, and pressures great enough to crush a human skull! More than 500 different organisms have been discovered near deep sea vents. These include strange and wonderful creatures such as giant clams, furry "yeti" crabs, vent crabs, and soft-bellied spaghetti worms. Scientists believe these creatures are similar to early life forms on Earth.

did you know? EVEN THOUGH VENTS CAN REACH TEMPERATURES OF 752°F (400°C), HIGH PRESSURE KEEPS THE WATER FROM BOILING.

CORAL REEFS

Coral reefs are often called "rain forests of the oceans" because of the huge number of sea creatures that live there. The most essential inhabitant in a coral reef, however, is the coral. Reefs are formed by corals that live in groups, called *colonies*. A coral's body is a small, round, pouchlike sac called a *polyp*. The bottom of a polyp is attached to a surface, and the top consists of a mouth and tentacles. Some polyps are the size of a pinhead, while others are a foot (about 30 cm) wide. The coral polyp uses calcium from seawater to make a hard limestone cup to live in. After the coral dies, other corals build their homes on top of it. Millions of hard cups together form a coral reef.

Mount Otemanu rises in the center of the island.

▲ COLORFUL CORALS

Inside a coral polyp lives a special kind of one-celled algae. The algae use photosynthesis to make nutrients, which the coral shares. The coral, in turn, provides a safe place for the algae to live. These algae give corals their color. If the algae die, the corals turn white, a process called *coral bleaching*. Disease, pollution, and increased water temperature can all cause coral bleaching.

A SOUTH PACIFIC REEF ▶

This coral reef near the island of Bora Bora formed when coral larvae attached themselves to the submerged edges of an island volcano. Over time, the reef grew outward and upward and formed what is called an *atoll*, a ringed reef around the island. Atolls, along with other types of reefs, need warm water and sunlight to grow.

HOW CORAL REEFS FORM ▶

Coral reefs are formed from the skeletons of generation after generation of coral polyps. Most reefs are 5,000 to 10,000 years old. The sedimentary rock known as limestone can form from coral skeletons that are compacted to form rock. People use limestone to make cement and to neutralize acids.

did you
know? ....

ALTHOUGH CORAL REEFS COVER ONLY 0.2 PERCENT OF THE OCEAN FLOOR, THEY CONTAIN MORE THAN 25 PERCENT OF ALL MARINE LIFE!

Layers of lava and ash have built up from volcanic eruptions.

Living corals grow at or near the surface.

Vegetation grows on top of nonliving coral skeletons.

Corals grow in water that is warm, salty, shallow, and clear.

An edge of the reef

An atoll is a circular ring of coral reef that surrounds a volcanic island.

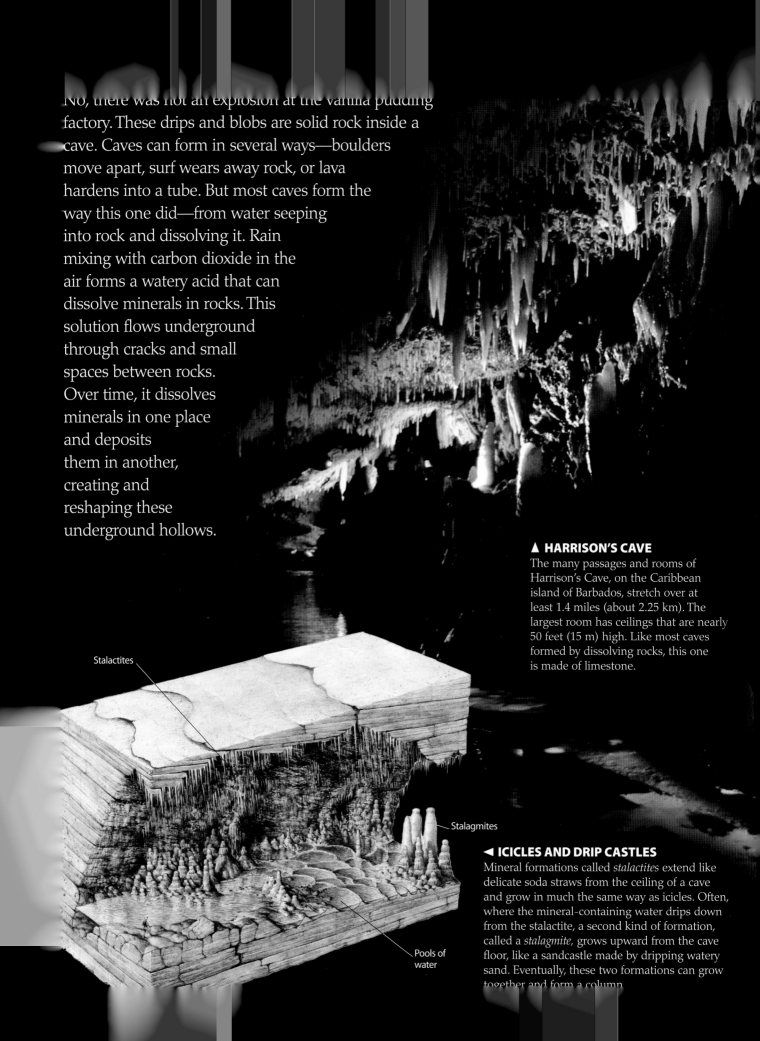

No, there was not an explosion at the vanilla pudding factory. These drips and blobs are solid rock inside a cave. Caves can form in several ways—boulders move apart, surf wears away rock, or lava hardens into a tube. But most caves form the way this one did—from water seeping into rock and dissolving it. Rain mixing with carbon dioxide in the air forms a watery acid that can dissolve minerals in rocks. This solution flows underground through cracks and small spaces between rocks. Over time, it dissolves minerals in one place and deposits them in another, creating and reshaping these underground hollows.

▲ HARRISON'S CAVE
The many passages and rooms of Harrison's Cave, on the Caribbean island of Barbados, stretch over at least 1.4 miles (about 2.25 km). The largest room has ceilings that are nearly 50 feet (15 m) high. Like most caves formed by dissolving rocks, this one is made of limestone.

Stalactites

Stalagmites

Pools of water

◄ ICICLES AND DRIP CASTLES
Mineral formations called *stalactites* extend like delicate soda straws from the ceiling of a cave and grow in much the same way as icicles. Often, where the mineral-containing water drips down from the stalactite, a second kind of formation, called a *stalagmite,* grows upward from the cave floor, like a sandcastle made by dripping watery sand. Eventually, these two formations can grow together and form a column.

This close-up of a stalactite shows how a mineral solution drips down from it.

▲ LIMESTONE CAVES

The mineral in limestone that is dissolved when caves form is called *calcite*. It is the chalky remains of organisms such as clams and corals that were deposited in ancient oceans. Landscapes that include hollows from dissolved rock, such as in many parts of Florida, are known as *karsts*.

did you know? IT CAN TAKE MORE THAN 100 YEARS FOR A STALACTITE TO GROW JUST HALF AN INCH!

GEODES

Giant crystals tower overhead. Sparkling crystals cover the walls, ceiling, and floor, surrounding you in glittering color. You are inside a geode! A geode is a hollow rock lined with mineral crystals. A few geodes are big enough to walk into, but most are small enough to hold in your hand. A geode forms in an empty space or pocket within a rock. Hot, mineral-rich water seeps through cracks in igneous rock, for example, depositing layers of microcrystalline minerals on the walls of the opening. These crystals are so tiny you can see them only under a microscope. But as the rock and water cool, larger crystals grow. Geodes can also form in sedimentary rocks, such as limestone. Sometimes the minerals in a geode are replaced by another mineral. Scientists can tell if this has happened because the new mineral has a different crystal shape!

Part of a geode's beauty comes from the layers of different types of minerals that have crystallized inside it.

Traces of iron turn plain quartz into purple amethyst. The purple is often darker near the crystals' tips.

AN ISLAND OF GEODES ▼

Geodes often form within rock that is softer than the minerals that make the geode. When the softer rock weathers away, the geodes remain. Here, on the island of Socotra, in the Indian Ocean, the weathering of soft limestone on this plateau has left behind a field of hard, round geodes.

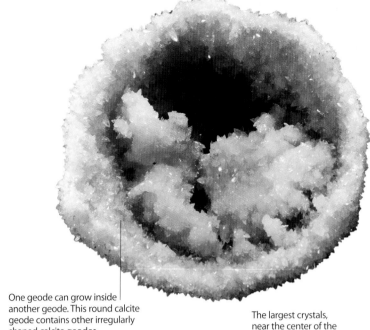

did you know?

ONE OF THE LARGEST GEODES IN THE WORLD WAS DISCOVERED IN 1999. ITS INNER LENGTH IS 26 FEET LONG (ALMOST 8 M) WITH CRYSTALS UP TO 6.5 FEET (ALMOST 2 M) LONG.

HIDDEN TREASURES ▼

Opening a geode reveals wonders that have been hidden from view until that very moment. You can crack geodes open with a hammer and chisel or cut them with a special rock saw to reveal more detail. You can also place a geode inside a sock and gently tap it with a hammer. So what's inside? Quartz is the most common geode-forming mineral. In a geode, it can form smooth, milky, banded agate or large, glassy amethyst crystals—or both. Calcite is another common geode-forming mineral. Some geodes contain multiple minerals. For example, calcite crystals can grow on top of quartz, or brass-colored metallic crystals of pyrite can be tucked among the larger crystals. If you have a large geode, try rattling it gently before opening it. You may hear loose crystals inside.

One geode can grow inside another geode. This round calcite geode contains other irregularly shaped calcite geodes.

The largest crystals, near the center of the geode, were the last to form and took the longest.

The crystals in this band of blue agate are submicroscopic— too small to be seen with an ordinary microscope.

MARBLE QUARRIES

It makes up the bone-colored bricks in the dome of the Taj Mahal. It gives Michelangelo's statue *David* its smooth, skinlike texture. Even a few toy marbles were made out of—you guessed it—marble! Whether part of an ancient work of art or a kitchen countertop, marble must first come out of the ground—at a marble quarry. A quarry is an open pit where stoneworkers cut rock out of the walls. Different quarries all over the world produce different colors and qualities of marble. Marble forms from a chalky rock—limestone. Limestone forms when sediment that is deposited in layers hardens. Limestone is dull, while marble has tiny crystals and can be polished. How does one type of rock turn into another? This process, called *metamorphism*, is a little like pressing with a spatula on a grilled cheese sandwich in a frying pan. Underground, heat and pressure cause melting and chemical changes in rocks that cannot be undone.

As the magma slowly cools, it forms igneous rock, such as granite.

Sandstone changes to quartzite.

Limestone changes to marble

Mudstone changes to slate.

A batholith is a large mass of magma that pushes its way into upper layers of rock.

▲ CHANGING ROCK

A large mass of slowly cooling liquid rock pushes up through layers of sedimentary rock beneath Earth's surface. This movement creates enough heat and pressure to melt the nearby layers. The metamorphic rocks that form have different colors, mineral grain sizes, and hardness than the sedimentary rocks they used to be.

THE WORLD'S MOST PERFECT MARBLE?

Marble comes in many colors: green, red, pink, blue, and even black. Thick veins of contrasting minerals can give it different textures. The different colors and textures come from mineral impurities that arise when the marble forms. A lack of impurities results in uniform white marble. The large block of marble that Michelangelo eventually carved into *David* came from the marble quarry of Carrara, located in what is now northwestern Italy. This quarry was famous during the Renaissance and still is today for its pure, dazzlingly white marble.

Sawing off flat blocks rather than uneven chunks makes the marble easier to transport, stack, and use in a wide variety of applications.

A worker checks the marble to ensure it has no faults or staining minerals.

MICHELANGELO'S *DAVID* ▼

Michelangelo was an Italian Renaissance artist most famous for his religious paintings on the ceiling of the Sistine Chapel in Vatican City and for this marble sculpture, *David*, located in Florence. Michelangelo carved *David* from a single block of Carrara marble. He completed the sculpture in 1504, at the age of 29. However, he was not the first artist to tackle the job. Other Florentine artists had already tried sculpting the same block of brilliant white marble decades before.

did you know?
AT 17 FEET (5.18 METERS), ABOUT 6 TONS, AND MORE THAN 500 YEARS OF AGE, *DAVID*'S ANKLES SHOW SIGNS OF STRESS.

Michelangelo imagined that when he carved a piece of marble, he was "freeing" the sculpture "imprisoned" in the stone.

EARTHQUAKES

What causes Earth to shake? Earth's crust is made of about twelve blocks of rock, called *tectonic plates*, sitting on a layer of hot molten rock. Most earthquakes occur where two plates meet. Pressure builds up as the plates try to slide under, over, or past each other. At some point, the plates move into a position that results in an earthquake. Some quakes are so mild that they can't be felt, and others shake the ground violently, destroying roads and buildings. The vibrations, called *seismic waves*, travel both on and below Earth's surface. The type of area they travel through influences how much destruction the waves cause.

HOW BIG WAS THAT QUAKE? ▼

The Richter scale records the magnitude of seismic waves. People usually don't feel earthquakes of 2.0 or less. Each whole-number increase indicates a tenfold increase in magnitude. A 5.0 is moderate, while a 6.0 is 10 times larger. Great earthquakes, of 8.0 or above, occur somewhere on Earth about once a year. Another scale, called the *Mercalli scale,* uses Roman numerals to rank earthquakes by how much damage they cause.

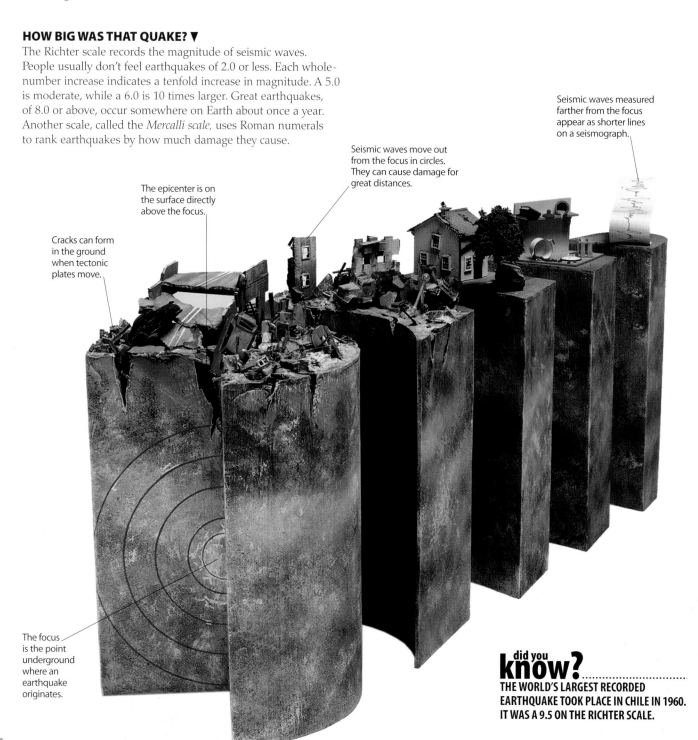

Seismic waves measured farther from the focus appear as shorter lines on a seismograph.

Seismic waves move out from the focus in circles. They can cause damage for great distances.

The epicenter is on the surface directly above the focus.

Cracks can form in the ground when tectonic plates move.

The focus is the point underground where an earthquake originates.

KOBE EARTHQUAKE ▼

In 1995, an earthquake of magnitude 7.2 on the Richter scale struck Kobe, Japan. The strong ground motions caused this expressway to collapse. Hundreds of thousands of buildings and homes were destroyed, and thousands of people were killed. The quake was a shindo 7 on a Japanese intensity scale that measures the degree of destruction from 0 to 7. Kobe was rebuilt with earthquake-resistant buildings and roads.

SEISMOGRAPHS MEASURE GROUND MOVEMENT ▲

An instrument called a *seismograph* records the seismic waves sent out by earthquakes. A pen makes a zigzag line when the ground under it moves. The bigger the movement sensed, the taller the line.

AFAR TRIANGLE

Blistering desert heat, miles of cracked earth spewing sulfur and lava, constant earthquakes, and almost no water— you have come to the Afar Triangle. This wedge of land, about the size of Nebraska, lies where Ethiopia borders the mouth of the Red Sea. Underneath the triangle, three giant pieces of Earth's crust meet in what is called a *triple junction*. The pieces, called *tectonic plates*, are pulling away from each other, stretching and thinning Earth's crust. Along the edges of the plates, volcanoes erupt. As the three plates drift apart, the land between the plates sinks. Some areas of the Afar Triangle are already more than 300 feet (100 m) below sea level. That is about as tall as a 30-story building! That's why many geologists call this area the Afar Depression.

SPLITTING UP ▼

The Afar Triangle is part of the East African Rift System, one of the largest systems of faults, or splits, in Earth's crust. Rifts are valleys that form when plates move apart. Over millions of years, one rift separated Africa and the Arabian Peninsula. Then the Red Sea filled in the gap. The rift forming in the Afar Triangle extends south beneath several East African countries. It could one day separate those countries from the rest of the continent.

Pools of sticky mud are all that remain after it rains in the Afar region, where one river barely supports the people who live along it.

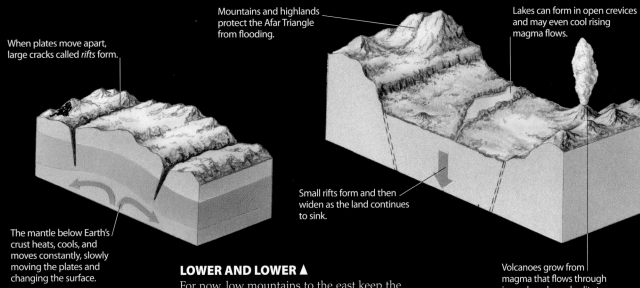

When plates move apart, large cracks called *rifts* form.

Mountains and highlands protect the Afar Triangle from flooding.

Lakes can form in open crevices and may even cool rising magma flows.

The mantle below Earth's crust heats, cools, and moves constantly, slowly moving the plates and changing the surface.

Small rifts form and then widen as the land continues to sink.

Volcanoes grow from magma that flows through jagged cracks and splits to the surface.

LOWER AND LOWER ▲

For now, low mountains to the east keep the Red Sea from flooding into the Afar Triangle, but these mountains are wearing down over time. Scientists predict that seawater will one day cover the Afar region.

did you know?..............
SOME OF THE OLDEST HUMAN-LIKE FOSSILS—MORE THAN 3 MILLION YEARS OLD—WERE FOUND IN THE AFAR REGION.

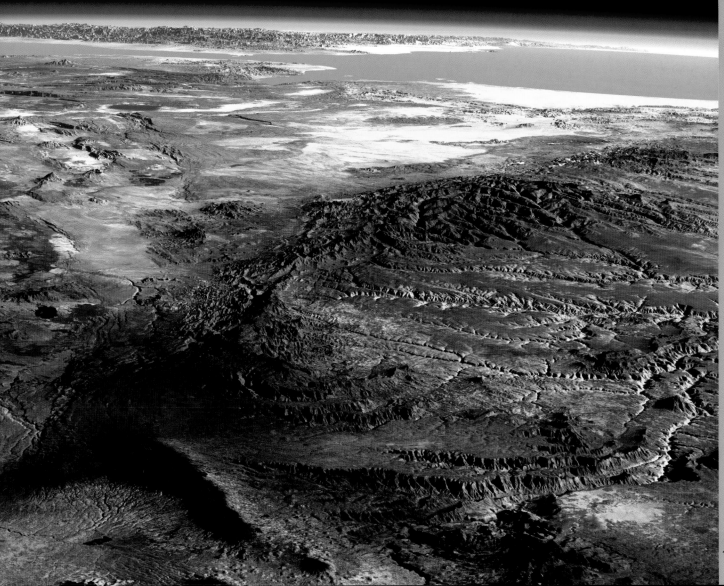

LANDSLIDES

Landslides are mass movements of earth, rock, or debris down a slope. They are natural hazards that occur all over the world. Landslides can be small, or so big that you can photograph them from space! Some move slowly—a few inches a year. Others are fast and catastrophic, at speeds of more than 175 miles an hour (about 281 km/h). These mass movements of earth are triggered by natural events such as earthquakes, rainstorms, volcanic activity, or wildfires. They can also be caused by human activities such as road building, flooding, or mining. Landslides can be very destructive. In 1970, a landslide triggered by an earthquake in Peru killed more than 18,000 people and destroyed two towns near Mount Huascarán. They can also reshape the landscape. For example, the huge landslide that accompanied the eruption of Mount St. Helens in the state of Washington in 1980 changed the shape of the mountain and the course of rivers.

did you know? ....
THE LARGEST LANDSLIDE IN RECENT HISTORY WAS TRIGGERED BY THE 1980 ERUPTION OF MOUNT ST. HELENS IN WASHINGTON STATE. IT WAS 14 MILES LONG (ALMOST 23 KM).

LANDSLIDE IN GUATEMALA ▼
This spectacular landslide occurred in Guatemala in January 2009. Officials believe this landslide was nearly a mile (1.6 km) wide! Millions of pounds of rock, earth, and mud tumbled down a mountainside, burying part of a road and killing at least 33 people. Geologists believe this landslide was triggered by a fault that runs through the area. Faults are cracks in Earth's crust that separate adjacent surfaces, making the surrounding area unstable.

TYPES OF LANDSLIDES ▼
There are many different types of landslides, but all happen when a weakened part of earth separates from a more stable underlying material. Rocks can fall or topple, soil can slide and spread, and mud can flow. For example, soggy soil can weaken and then move downhill or "slump." This image shows how this type of landslide wiped out part of a road in Portugal.

Rock debris buried part of the road.

A pile of rock debris that collects at the bottom of a landslide is called a *talus*.

◄ **SEEN FROM SPACE**

A NASA satellite captured this image of a massive landslide that occurred in China's Chongqing region in 2009. A mountainside collapsed and filled the valley below with 420 million cubic feet (almost 12 million m³) of rocky debris and earth. The landslide buried houses, power lines, and part of an iron ore mine, killing residents and miners.

Debris field

One of the two roads that were partially buried

The very end of the landslide is called the *toe*.

This long, clifflike edge is called a *scarp*; it marks a place from which land broke away.

KILAUEA

Kilauea in Hawaii has been active for between 300,000 and 600,000 years, making it one of the most active volcanoes in the world. A volcano does not have to be erupting to be considered active— an active volcano is simply capable of venting lava, ash, vapor, and gases. Kilauea is located on the Pacific plate, one of Earth's tectonic plates. It is situated directly above a hotspot, a column of magma that reaches Earth's crust and forms a vent. Kilauea began as an undersea vent, erupting with lava repeatedly until it emerged from the ocean as an island between 50,000 and 100,000 years ago. Usually volcanoes that form above a hotspot die as the tectonic plate moves away from the column of magma. Most of the islands in the Hawaiian chain are dormant volcanoes that have moved away from the hotspot. Kilauea, however, remains above the hotspot—and active.

Trade winds carry water vapor, carbon dioxide, and sulfur dioxide to the coast, creating volcanic smog, called *vog*, that can affect air quality.

did you know? SINCE 1983, KILAUEA HAS PRODUCED ENOUGH LAVA TO PAVE A ROAD TO THE MOON FIVE TIMES.

KILAUEA'S ERUPTION AREAS

Kilauea erupts from three main areas: a caldera (crater) at the summit and two rift zones (fractures or cracks) located high up the volcano's sides. Lava flows into the caldera and cools, heightening the volcano. Lava that emerges from the rift zones creates ridges that extend outward from the summit. As it flows downhill, the lava cools, gradually building up the volcano's shieldlike form.

The caldera is about 3.7 miles (6 km) across.

Lava that erupts from Kilauea's cone flows through a system of lava tubes (closed channels formed by continuous lava flow) to the sea.

The most recent eruption at Kilauea has been ongoing since January 1983.

PRE-ERUPTION ▲
As magma rises to Earth's surface, tremors, earthquakes, and ground uplift occur in the vicinity of the volcano. Sulfur dioxide gas pressure builds and the summit of Kilauea inflates, like the top of a soda can that has been shaken.

ERUPTION STARTS ▲
The concentration of sulfur dioxide emitted at the summit increases and becomes hazardous to tourist and residential areas downwind. Summit vents exhibit a dull red glow from rising lava, and small streams of lava begin to flow.

ERUPTING ▲
Plumes of lava may rise up to about 1,000 feet (300 m) above the volcano's rim. Usually this lava flows down the volcano's lava tubes. Occasionally explosions at the upper rift zones or summit spew steam, lava, and rock fragments over the surrounding landscape.

LIFE RETURNS TO LAVA FIELDS ▲
Fern spores and seeds carried by the wind fall into cracks in lava fields. Plants that take root can reach fertile soil below the hardened lava.

LAVA

While you attend school each day or spend time with your friends, Earth is shifting and changing under your feet. You may not actually feel it, because the movement is so slow. But you hear about it whenever an earthquake or volcanic eruption makes the news. Magma—fiery-hot molten rock—flows beneath Earth's crust. Volcanoes form where intense heat and magma escape to Earth's surface, usually along the edges where tectonic plates meet. Magma that reaches Earth's surface is called *lava*. The temperature and viscosity of magma (how fluid it is) and the amount of dissolved gases in it affect how the lava will erupt. Some lava erupts with a violent explosion, sending rocks, dust, and ash into the air. Other lava forms a lava flow that pours out of a volcano.

Pumice forms when gas-filled, frothy lava explodes from a volcano and hardens. Pumice is a lightweight rock and floats on water.

Pahoehoe lava is smooth, often ropy lava that is common in lava flows.

WHEN LAVA COOLS ▲

As lava cools, it forms volcanic igneous rock, turning black, gray, or dark red. Volcanic igneous rock contains fine crystals and is often glassy. Lava that flows directly into the ocean can cool so fast it shatters into sand. Pillow lava forms when molten lava breaks through the thin wall of an underwater lava tube. The lava squeezes out like toothpaste, creates irregular tonguelike shapes, and quickly hardens.

did you know? ....
HAWAII'S BLACK SAND BEACHES WERE CREATED INSTANTANEOUSLY WHEN HOT LAVA SHATTERED AS IT REACHED THE SEA.

This lava flow occurred near Hawaii's Kilauea volcano in March 2007. The upper lava layer has cooled and hardened.

Lava drips, called *driblets*, can harden into many shapes.

Obsidian is a type of volcanic glass. It is composed of melted sand (the primary ingredient of glass).

LAVA TUBES ▼

Volcanic eruptions can last a long time, creating streams of lava that flow for several hours or days. Ongoing lava streams create flow channels on Earth's surface. When the outer edges of a channel cool and harden, the sides build up. A crust can form over the top of the channel, creating a lava tube. Lava that flows through lava tubes stays hot and fluid much longer than surface lava. When the eruption ends, the lava flows out of the tubes, leaving caves and tunnels, often large enough for people to explore.

LAVA ON THE MOVE ▼

Scientists identify lava types not only by how they erupt but by their silicon, oxygen, iron, and magnesium content. Common lava flows swiftly because it contains less silicon and is therefore thinner than lava that contains high amounts of silicon. Dissolved gases rise easily to the surface of thin lava, so eruptions are not explosive. Dissolved gases cannot easily rise through the silicon of thicker, slower-moving lava. Instead, the gases build up pressure, and when the gas bubbles finally reach the lava's surface, they explode.

The lower layer is still hot and flowing because the crust above helps hold in heat.

159

GEYSERS

What do you get when Mother Earth lets out a steaming burp? A geyser! A geyser is a hot spring that has eruptions. These eruptions send steam and boiling hot water into the air. There are only about 1,000 active geysers on Earth. They are so rare because they form only under very specific conditions. For a geyser to form, there must be a lot of water filling a system of watertight underground cracks. These pipelike cracks must be able to withstand great pressure. Most importantly, this water must be located near a very hot place—such as an underground pocket of melted rock, or magma, that feeds a volcano. Such heat from deep underground is called *geothermal energy*. In nature, geothermal energy powers geysers, many kinds of rock changes, and volcanoes. People use geothermal energy, too. Geothermal power plants are like human-made geysers. The hot steam that comes up can be used to power turbines that generate electricity.

OLD FAITHFUL ▼
This reliable geyser erupts every 65 to 92 minutes for a period of 1.5 to 5 minutes. Old Faithful is one of the most frequently erupting of the big geysers in Yellowstone National Park in Wyoming. It sends 3,700–8,400 gallons (about 14,000–31,800 L) of water into the air during each eruption.

Colorful mats of heat-loving bacteria thrive in the hot springs near geysers.

FLY GEYSER ▲
These colorful shapes look like plastic fountains you might see at an amusement park. They are actually rocky mounds deposited by a man-made geyser. In 1964, a company looking for geothermal energy drilled a test well in Nevada. The 200°F (93°C) water was not hot enough for their needs, but after they left, the water kept bubbling up from the ground. Over time, the hot water deposited minerals that built up around the openings in the ground. Various types of heat-loving algae give the rocks their color.

Old Faithful's column of water can shoot as high as 184 feet (about 56 m) in the air.

The rims of hot springs and cones of geysers are made up of deposits of dissolved rock, called *sinter*.

At the surface, the steam rises into the air, followed by the boiling water that has built up. Cooled water seeps back into the ground to begin the process once again.

Pipelike underground cracks

Water heated by hot rocks forms bubbles of water vapor, which can be trapped in narrow passageways.

Hot spring

HOW A GEYSER WORKS ▲

The boiling point of a substance increases with pressure. Water deep underground is at high pressure due to the weight of the water above. So this water must reach temperatures higher than 212°F (100°C) to boil. Once this water starts to boil, bubbles of water vapor travel up toward the surface. These bubbles get trapped in the narrow passageways. As more bubbles are trapped, the force on the water above increases until a small amount of water is pushed out of the geyser. Once this water is out of the way, there is less pressure on the water underneath. Less pressure means the water will boil at a lower temperature—one it has already reached. All of the water boils at once, sending steam and hot water erupting out of the geyser.

ISLANDS

On a globe, Earth's landmasses appear to have water all around them. So, are all landmasses islands? No. Islands are completely surrounded by water—but they are smaller than a continent. They also differ from continents in the way they form. Scientists believe that the continents were created by plate tectonics—the theory stating that fragments of Earth's crust and uppermost mantle shift or float on the rest of the mantle. Most islands, however, form in three main ways. Volcanic activity below the ocean floor caused oceanic islands, such as the Hawaiian Islands, to form and rise above sea level. Continental islands, such as Greenland and New Guinea, are parts of continental shelves. They became isolated when glacial ice melted, flooding and covering the land that connected them to the continent. Islands like the Maldives, located off the coast of India, arose from coral reefs. Over time, enough sand and dust accumulated on the reefs to form islands.

Many of the Rock Islands appear mushroom-like. They are made of easily dissolved limestone that is undercut at the waterline.

did you know?..................
KILAUEA, A VOLCANO ON THE ISLAND OF HAWAII, HAS BEEN ERUPTING NEARLY CONTINUOUSLY SINCE 1983.

THE ISLANDS OF PALAU
The Republic of Palau, an archipelago (group of islands) located near the Philippines, includes volcanic, coral, low limestone, and high limestone islands. Some of the islands are a combination of types. The Rock Islands (shown here) and other limestone islands formed when tectonic plates shifted. The shift pushed parts of ancient coral reefs and ocean floor above sea level.

HOW VOLCANIC ISLANDS FORM ▼

Volcanic islands form when oceanic plates collide and the edge of one plate subducts, or slides under another. The subducted edge melts, and the magma rises to form an island. Volcanic islands also form when oceanic plates move across hot spots in Earth's mantle, a process shown in the diagram below.

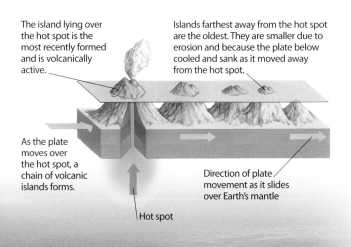

The island lying over the hot spot is the most recently formed and is volcanically active.

Islands farthest away from the hot spot are the oldest. They are smaller due to erosion and because the plate below cooled and sank as it moved away from the hot spot.

As the plate moves over the hot spot, a chain of volcanic islands forms.

Direction of plate movement as it slides over Earth's mantle

Hot spot

In the absence of a true soil layer, vegetation on the islands grows out of the loose limestone rock.

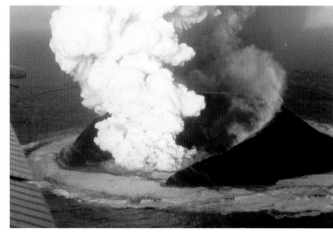

▲ THE BIRTH OF AN ISLAND

In 1963, undersea volcanic eruptions heaved up a new island from the ocean floor about 20 miles (32 km) south of Iceland. Named Surtsey, this island belongs to a volcanic system of islands and underwater cones that crosses east central Iceland.

▲ THE FORMATION OF SURTSEY

By the time volcanic eruptions stopped in 1967, Surtsey was 492 feet (150 m) above sea level and spanned about 1 square mile (almost 3 sq km). The ocean eroded parts of the island before its core solidified as rock.

▲ THE GREENING OF SURTSEY

The general public cannot visit Surtsey, so plants and animals are able to colonize there without threat. Ocean currents, wind, and birds carry seeds and organisms there. Scientists can study the natural progression of colonization and observe succession, the changes in species populations.

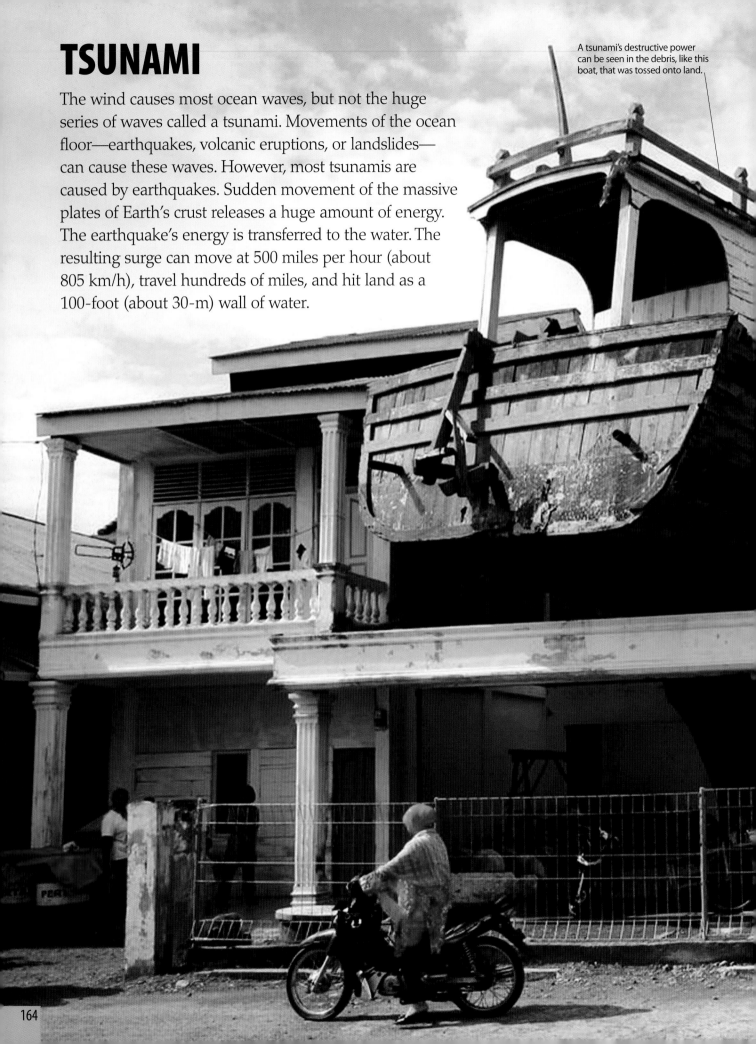

TSUNAMI

The wind causes most ocean waves, but not the huge series of waves called a tsunami. Movements of the ocean floor—earthquakes, volcanic eruptions, or landslides—can cause these waves. However, most tsunamis are caused by earthquakes. Sudden movement of the massive plates of Earth's crust releases a huge amount of energy. The earthquake's energy is transferred to the water. The resulting surge can move at 500 miles per hour (about 805 km/h), travel hundreds of miles, and hit land as a 100-foot (about 30-m) wall of water.

A tsunami's destructive power can be seen in the debris, like this boat, that was tossed onto land.

1. Where Earth's plates meet on the seafloor, one plate is pushed up.

2. The water above the uplifted seafloor is suddenly pushed up.

3. The rising water causes waves in the deep ocean.

4. As the waves move into shallower water, their wavelength shortens and the wave height increases.

5. Waves become tall and destructive in shallow water.

◄ THE POWER OF WATER

On December 26, 2004, a tsunami in the Indian Ocean killed more than 200,000 people and destroyed thousands of buildings. It dropped this fishing boat on top of this house on the island of Sumatra in Indonesia. The earthquake that caused the tsunami registered 9.0 on the Richter scale and occurred about 150 miles (about 241 km) away from where the wave struck land. Huge waves also reached the coast of Africa, more than 3,000 miles (more than 4,800 km) away.

▲ HOW A TSUNAMI FORMS

Tsunamis travel quickly through deep water. The waves move in all directions from the earthquake's center. In deep water, they are seldom larger than normal waves and may not be noticed by ships at sea. Tsunami waves slow down as they run into the shallower water closer to land. The wave is compressed, forcing more water into each peak and trough. This causes the wave to grow dramatically taller.

did you know?.....................

A TSUNAMI CAN TRAVEL ACROSS THE PACIFIC OCEAN IN A SINGLE DAY.

FLOODS

Although water is necessary for all life, a flood is too much of a good thing. Floods most often occur because more rain falls than an area can absorb in a given period of time. This can cause landslides, broken dams, and rising rivers. When rivers rise slowly, people may have time to leave the area before water overflows the banks. When torrential rain quickly sweeps into an area, it can cause what is called a *flash flood*. Because these floods happen too quickly for people to get to higher ground, flash floods can cause many deaths. Tsunamis, hurricanes, and broken dams can create dangerous waves, storm surges, or moving walls of water that overrun everything in their path. Entire drainage systems can overflow, especially in urban areas where there is not enough open land to soak up the water. Over the last century, the highest death toll—several million people—from a natural disaster was from a 1931 flood in China.

SOUTHERN CHINA, 2008 ▼
In June of 2008, large areas of southern China experienced day after day of heavy rainfall. Because the water level rose slowly, many people were able to evacuate. The floods caused landslides; destroyed homes, roads, and crops; and displaced more than a million people.

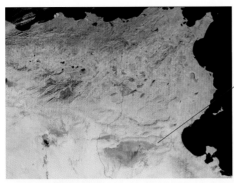

On January 4, 2003, Tunisia is at the start of the winter rainy season.

Depressions in the land, called *salt pans*, collect water, until the water evaporates.

Giant piers contain the machines that raise and lower the barrier gates that are between the piers.

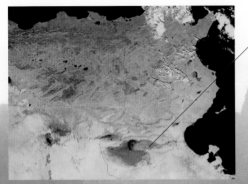

On January 19, 2003, more areas of blue and gray show flooding.

The darkening and spreading of the blue colors show that the water is deeper and that the salt-pan lakes have grown.

◀ NORTHERN AFRICA, 2003

Extreme conditions—cold temperatures, very heavy rains, and snow in mountainous areas—caused floods in northern Tunisia and Algeria during the winter of 2003. NASA photographs taken about two weeks apart show more snow and water, in shades of blue. Flooding drove 3,000 people from their homes.

▲ THE THAMES BARRIER

The Thames Barrier is the largest movable flood barrier in the world. It protects the city of London from flooding. Normally, the barrier gates are lowered to allow the Thames River to flow and ship traffic to pass. During tidal surges, the barrier gates are raised to hold back water that travels up the river from the sea.

A survivor guides his craft atop water-clogged streets to get his passenger to safety.

did you know?
SOME ENGINEERS PROPOSE A "GIVE WATER SPACE" POLICY FOR THE NETHERLANDS AND NEW ORLEANS: A PLAN TO BUILD CANALS AND STORAGE AREAS FOR WATER TO FLOW INTO, NOT JUST DIKES AND LEVEES TO KEEP WATER OUT.

ATMOSPHERE

Earth's atmosphere has a big job. It's like bubble wrap, protecting the planet and the life on it from the harsh conditions of space. It filters out dangerous radiation, stops meteors, and helps transfer heat across the globe. Billions of years ago, volcanoes belched out gases such as carbon dioxide, nitrogen, and water vapor. Some of those gases were held in by Earth's gravity. Many biochemical processes—cloud formation, rain, rock formation, and photosynthesis—eventually added oxygen to the mix.

Now oxygen makes up 21 percent of the atmosphere. Oxygen, nitrogen, and traces of carbon dioxide and water vapor form an atmosphere that provides the materials for sustaining life on Earth.

UP TO THIN AIR
The layers of the atmosphere differ from one another in the number of gas particles they contain. The closer a layer is to Earth, the denser it is, because more gas particles are held by gravity. The troposphere and the stratosphere, together extending just 30 miles (50 km) above Earth's surface, contain 99 percent of the gases in the atmosphere. The air becomes increasingly thinner in the mesosphere, thermosphere, and exosphere.

The lower atmosphere holds most of the world's water vapor, giving rise to clouds and severe storms.

Thunderstorms can send special lightning—red rings, called *sprites*, and blue streaks, called *blue jets*—into the upper atmosphere.

HOT OR COLD UP THERE?

Each of the first three layers of the atmosphere is topped by an area called a *pause*, where temperatures change. As you climb to the tropopause, the top of the troposphere, the temperature drops to -60°F (-51°C). The stratosphere warms with altitude, to about 5°F (-15°C), as ozone forms a layer that absorbs the sun's UV radiation. The mesosphere has few particles to absorb solar radiation. It gets colder as you go up, reaching -184°F (-120°C). The thermosphere has even fewer particles, but they are closer to the sun and can heat up to 3,600°F (2,000°C).

did you know? SCORCHING PARTICLES IN THE THERMOSPHERE ARE SO FAR APART THAT THE AIR FEELS COOL.

Upper thermosphere, where air is so thin that it is often considered part of outer space

Lower thermosphere, where the space shuttle flies and auroras happen

Mesosphere, where most meteors burn up as shooting stars

Stratosphere, where commercial jets fly in the stable air layers

Troposphere, where most weather forms and small airplanes fly

Land and sea surfaces interact with the atmosphere.

Bedrock within Earth's crust separates magma from the surface.

High-energy gases dissolved in magma can help eject dust from erupting volcanoes even into Earth's stratosphere.

AURORA BOREALIS

You see a strange, glowing light in the corner of the night sky. The mysterious light grows into a swirling cloud of green and red that fills the sky above. Then, within hours, it fades back into darkness. You have just seen an aurora! An aurora is a natural light display seen at night in the polar regions of Earth. Auroras happen when charged particles from the sun reach the magnetic field that surrounds Earth and are trapped. Many of these trapped particles move toward Earth's magnetic poles. There, they can run into gas molecules in the atmosphere. These collisions give off light energy, producing an aurora. In the Northern Hemisphere these strange and beautiful lights are called the *aurora borealis,* or the northern lights. In the Southern Hemisphere, they are called the *aurora australis,* or the southern lights.

NORTHERN LIGHTS ▶

Most auroras occur about 60 miles (100 km) above Earth, in the thermosphere layer of the atmosphere, though they can occur 10 times higher. Auroras can have many different colors of light, caused by the different types of gas molecules in the atmosphere. Oxygen most often makes green light, the most common color of an aurora. Blue light is given off when the charged particles collide with nitrogen. Some of the light given off is ultraviolet light, which we cannot see.

know?
did you

THE COLLISIONS THAT CAUSE AURORAS ALSO TAKE PLACE DURING THE DAYTIME, BUT THEY ARE NOT BRIGHT ENOUGH TO BE SEEN.

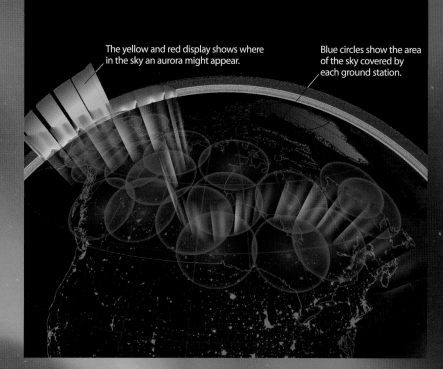

The yellow and red display shows where in the sky an aurora might appear.

Blue circles show the area of the sky covered by each ground station.

Aurora lights that occur very high in the sky can appear red or purple.

TRACKING AURORAS ▲
Sometimes an aurora will brighten, break up into smaller parts, and dance across the sky as it changes color. The cause of this special type of dancing aurora is unknown. To solve this mystery, NASA scientists will use data from probes launched into space and cameras on the ground. This image shows how the ground stations might detect an aurora.

Aurora borealis makes the sky appear green in Manitoba, Canada.

Some aurora displays can spread thousands of miles across the sky.

WEATHER FRONTS

Air masses are like sumo wrestlers belly bumping in the atmosphere. Suppose a huge body of cold, dry air moves toward a huge body of warm, wet air. The cold air is more dense and slides under the lighter warm air. The warm air rises, cools off, and may form clouds that drop rain. The greater the difference in the temperature and humidity of air masses, the more intense the weather will be when they meet. That's why the boundary between air masses is called a weather *front*, the place where battles take place. Air masses originate in areas called *source regions*. When slow-moving air hangs over these large, mostly uniform stretches of land or water, the air takes on the characteristics of the land below. Dry, or continental, air masses form over land; moist, or maritime, air masses form over oceans. Cold, or polar, air masses form over polar regions; warm, or tropical, air masses, form near the equator.

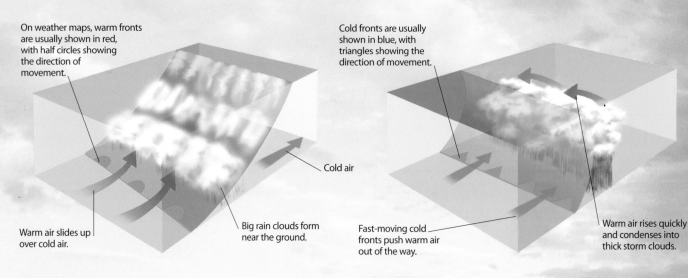

On weather maps, warm fronts are usually shown in red, with half circles showing the direction of movement.

Cold fronts are usually shown in blue, with triangles showing the direction of movement.

Cold air

Warm air slides up over cold air.

Big rain clouds form near the ground.

Fast-moving cold fronts push warm air out of the way.

Warm air rises quickly and condenses into thick storm clouds.

WARM FRONT ▲

The diagram above shows a warm air mass, in red, moving toward a cold air mass, in blue. The leading edge of the warm air mass is a warm front. The warm air is lighter, so it slides slowly up on top of the cold air. Water vapor in the warm air condenses as the air rises and cools, so clouds form. These clouds may dump heavy rain.

COLD FRONT ▲

Above, the blue cold air mass is moving toward the red warm air mass. The cold air is heavy and usually moves faster, pushing the warm air out of the way. If the warm air is also humid, its water vapor may condense and form thunderstorms. Typically, the cold air that passes through after the storms is drier.

did you know?..............
AIR MASSES TYPICALLY COVER HUNDREDS OF THOUSANDS OF SQUARE MILES (MILLIONS OF KM²).

SQUALL LINE: FAST AND FURIOUS ▶

A sudden gusty wind that usually comes with rain is called a *squall*. Squall lines like this one form along fast-moving cold fronts. The row of dark clouds marks the boundary where the cool air mass is pushing up a warm, humid air mass. Until now, this was a good beach day! But severe thunderstorms can form when the warm humid air starts to cool as it rises. Then the bad weather approaches along the advancing front.

GUST FRONT ▶

Unstable atmospheric conditions can have dramatic results. This curved cloud at the edge of a thunderstorm shows the location of a gust front. A gust front is the leading edge of gust winds that are formed by the strong downward currents of air in a thunderstorm. Some gust fronts are strong enough to damage buildings and knock down trees and power lines.

THUNDERSTORMS

Thunderstorms are nature's display of fireworks. They produce lightning and thunder, and are usually accompanied by rain or hail and wind. Beautiful and powerful, thunderstorms can also be deadly. Clouds form as moist air rises from Earth's warm surface. As the air cools down, the clouds fill with millions of particles of ice. Those particles collide with each other as the wind moves them up and down inside the clouds. This collision of particles is what builds up electrical charges. As negatively charged particles are attracted to areas of positive charge, they produce a large spark, which is lightning. Some thunderclouds build a negative electrical charge at the bottom of the cloud. This causes Earth's surface to become positively charged, through what's called *induction*. Negatively charged particles on Earth's surface are repelled by the like charges at the bottom of the cloud, so they move away. This leaves Earth's surface with a positive charge. When you see a lightning bolt strike the ground, you are actually seeing negative charges, or electrons, moving from the clouds to the ground. The positive ground charge tends to concentrate on elevated areas such as antennas, trees, or hills. Standing in such locations during a thunderstorm is very dangerous—you are an easy target!

A discharge of built-up energy produces lightning.

The longest recorded lightning bolt was 118 miles (about 190 km) long.

You see a powerful flash of light when the negative charge of lightning meets the positive charge of Earth's surface.

◄ THUNDER

The sound of thunder in the distance warns you that a storm may be heading your way. If you hear thunder, look for cover! When lightning flashes across the sky, you hear thunder a few seconds after you see the light. Light travels faster than sound, so the light of the bolt reaches your eyes before the sound reaches your ears. What produces the sound of thunder? Lightning heats the surrounding air, sometimes by as much as 50,000°F (about 33,000°C). That's almost 5 times the temperature of the sun's surface! This hot air expands very fast, causing a shock wave to radiate in all directions. The shock wave travels as a sound wave that makes the sound of thunder. Thunder makes a rumbling sound because you are hearing sound waves that radiate from different parts of the zigzag lightning bolt.

Lightning is so hot it can melt sand and turn it into amazing glass tubes called *fulgurites*. These tubes form in one second and take the shape of the lightning as it hits the sand.

▲ THUNDERSTORMS ARE BENEFICIAL

Thunderstorms are beneficial to Earth in many ways. Lightning produces nitrogen oxides in the atmosphere that react with other chemicals and sunlight to produce ozone—the gas that protects Earth from ultraviolet radiation. Thunderstorms also help plants. Plants can't absorb nitrogen through their leaves but they can absorb it dissolved in water. Lightning helps nitrogen dissolve in water, which then gets into the soil. Finally, thunderstorms help maintain Earth's electrical balance. Electrons from Earth's surface are constantly flowing upward, and thunderstorms transfer electrons back to Earth.

did you know? LIGHTNING FLASHES SOMEWHERE IN THE WORLD MORE THAN 3 MILLION TIMES PER DAY!

RAINBOWS

Whether they form while you are washing the car or during a cloudburst, rainbows happen because of light energy interacting with matter. Rainbows form from sunlight that is both reflected and refracted by water droplets suspended in the atmosphere. To reflect means to bounce light, such as when light hits a mirror. To refract means to bend light. Instead of traveling along a straight path, a beam of light bends or moves off at an angle from the object it strikes. Refraction happens when white light strikes a prism or a raindrop. White light is made up of many colors of light, called *wavelengths*. A rainbow forms because water refracts the different wavelengths at slightly different angles and separates the colors. Violet light bends more than red light. The farther the light travels from where it refracts, the more spread out the colors of the rainbow appear in the sky.

◄ THE MAGIC NUMBER

Rainbows form when sunlight enters and leaves water droplets at a 42-degree angle. As long as that condition is met, even spray from a waterfall can form a rainbow. For this reason, the highest point at which a rainbow can form is at a 42-degree angle above the horizon. If the sun is higher than that, a rainbow cannot form.

A RAINBOW'S SHAPE AND COLOR ▲

Rainbows are arc-shaped because water droplets are round and the inside surface that reflects the light is curved. At sunset, rainbows are semicircular. When the sun is higher, the arc is smaller. A rainbow's color intensity is affected by the size of the water droplets. Large droplets produce bright, well-defined rainbows. Tiny droplets form overlapping color bands that appear almost white.

UNDERSTANDING ALL THE ANGLES ▼

Sunlight, in the form of white light, refracts as it passes from the atmosphere into a water droplet. When the light strikes the back of the droplet, it reflects at an angle, and then refracts again as it passes back out of the droplet. Different colors refract at different angles because they travel at different speeds when they pass through water. The sunlight separates into the visible color spectrum on their way out. Rainbows can form a complete circle, because a circular droplet creates a circular reflection—but the horizon cuts the circle in half.

Raindrop

Red light has the longest wavelength and violet light has the shortest.

did you know?

BRIGHT MOONLIGHT CAN CAUSE A "MOONBOW," OR LUNAR RAINBOW. IT IS HARD TO SEE A MOONBOW'S COLORS, HOWEVER, BECAUSE THE REFRACTED LIGHT IS DIM.

PREDICTING HURRICANES

Hurricanes are one of nature's most destructive storms. Long ago, people had no way of knowing when a hurricane was approaching. In 1900, a hurricane struck Galveston, Texas, and 6,000 or more people died when the island was flooded. Such a loss is unlikely today, because forecasters can predict 5 days in advance how strong hurricanes will be and where they might make landfall. How do forecasters know so much about hurricanes? They use modern equipment like satellites, airplanes, radar, ocean buoys, and sophisticated computer modeling systems. Satellites can see the ocean where there are few ships. They can track cloud formations and ocean temperatures. Doppler radar can monitor wind data and precipitation levels. Ocean buoys send back data on air and water temperature, wave height, and wind speed. Airplanes drop tiny weather stations into the storm to get up-to-date information. Complicated computer programs analyze all the data to predict hurricane behavior.

did you know?
THE MOST DESTRUCTIVE PORTION OF A HURRICANE IS FOUND IN THE EYE WALL—WHICH BORDERS THE CALMEST PART OF THE HURRICANE, THE EYE.

LOOKING INTO HURRICANE IVAN
NASA's Tropical Rainfall Measuring Mission (TRMM) satellite, originally designed to measure rainfall, allows scientists to see rain patterns inside hurricanes. With TRMM, meteorologists can better forecast hurricane intensity. Hurricane Ivan (shown in the large background image) was one of the worst Atlantic hurricanes ever recorded. Ivan caused enormous damage and spawned 117 tornadoes in the United States. Because storm forecasts were so accurate, however, fewer than 100 people died.

❶ HURRICANE RITA ENTERS THE GULF OF MEXICO Ocean regions that have sea surface temperatures of 82°F (almost 28°C) or more (indicated as red and orange areas) are warm enough to form a hurricane.

❷ RITA GATHERS STRENGTH The hurricane winds are strengthened by the heat energy from the warm ocean. Sea surface temperatures cool as the hurricane passes.

❸ RITA MAKES LANDFALL Rita makes landfall on the Texas-Louisiana border. Because the ocean no longer supplies energy, Rita quickly downgrades from an intense hurricane to a tropical storm.

The National Oceanic and Atmospheric Administration (NOAA) monitors the Western Hemisphere with satellites similar to this one.

STORM CHASERS ▲

Some experienced meteorologists pursue severe weather events in trucks called *Dopplers on Wheels* (DOW). They use radar to collect data from inside storm cells. This up-to-the-minute, localized storm information is added to other collected data to help scientists forecast the weather more accurately. DOWs have shed light on how hurricanes intensify. Here, a DOW collects data as Hurricane Frances approaches Florida in 2004.

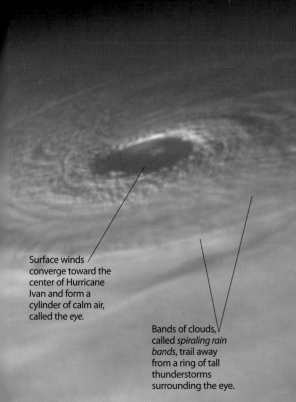

Surface winds converge toward the center of Hurricane Ivan and form a cylinder of calm air, called the *eye*.

Bands of clouds, called *spiraling rain bands*, trail away from a ring of tall thunderstorms surrounding the eye.

SATELLITE WEATHER OBSERVATION ▲

The United States uses stationary and polar-orbiting satellites to observe weather and other phenomena 24 hours a day. These satellites track fast-breaking storms and tornadoes in the country's interior and tropical storms in the Atlantic and Pacific oceans. This 3-D model made from a satellite image of Hurricane Wilma shows its eye and rings of moderate to intense rain. Red portions indicate areas of heaviest rainfall. At the time of this image, Wilma had sustained wind speeds of 150 miles per hour (about 241 km/h).

FOG

We can see fog, but we can't touch it. So what is it? Fog is a cloud that forms close to the ground. Water is continuously evaporating from Earth's surface, adding water vapor to the air. Water vapor is water in a gaseous state, and it's invisible. Air can become what is called *saturated*—it holds as much water vapor as possible—also referred to as a condition of 100 percent humidity. As air cools, some of the water vapor condenses into liquid water droplets. As these water droplets form, they may cling to particles in the air, such as dust, pollution, or salt. A low-lying patch of water droplets clinging to particles is called *fog*. Such an area much higher in the atmosphere is called a *cloud*. Fog is defined as a condition in which visibility is less than 0.6 miles (about 1 km). When visibility is greater than 0.6 miles, the condition is called mist.

◄ FOG OVER MONT ST. MICHEL
France's Mont St. Michel is a rocky island surrounded by tidal mud flats. Fog forms here on clear nights when the mud cools. The cool mud also cools the air above it. The water droplets condense onto salt particles suspended in the air to create ground-level fog that rarely moves. It usually disappears after the sun rises, because warm air evaporates the water droplets.

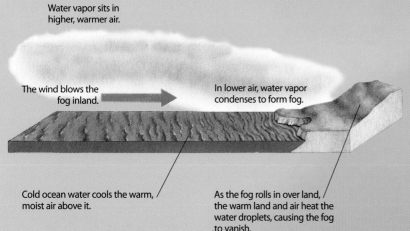

Water vapor sits in higher, warmer air.

The wind blows the fog inland.

In lower air, water vapor condenses to form fog.

Cold ocean water cools the warm, moist air above it.

As the fog rolls in over land, the warm land and air heat the water droplets, causing the fog to vanish.

◄ FOG ROLLS IN
Fog forms at sea when warm, moist air drifts over cold water. The water cools the air, and condensation takes place. Sea fog is "glued" together when condensed water attaches to salt particles tossed into the air by crashing waves. Salt is an unusual condensation particle. It will allow fog to form when the humidity is only 70 percent—that's less than complete saturation.

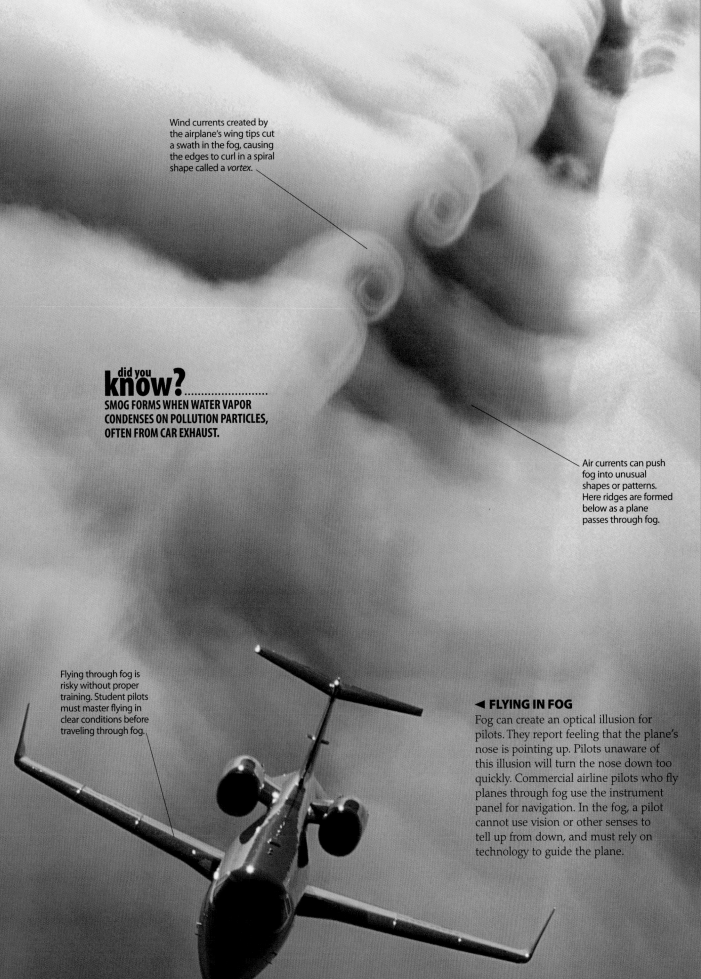

Wind currents created by the airplane's wing tips cut a swath in the fog, causing the edges to curl in a spiral shape called a *vortex*.

know? did you
....
SMOG FORMS WHEN WATER VAPOR CONDENSES ON POLLUTION PARTICLES, OFTEN FROM CAR EXHAUST.

Air currents can push fog into unusual shapes or patterns. Here ridges are formed below as a plane passes through fog.

Flying through fog is risky without proper training. Student pilots must master flying in clear conditions before traveling through fog.

◄ FLYING IN FOG
Fog can create an optical illusion for pilots. They report feeling that the plane's nose is pointing up. Pilots unaware of this illusion will turn the nose down too quickly. Commercial airline pilots who fly planes through fog use the instrument panel for navigation. In the fog, a pilot cannot use vision or other senses to tell up from down, and must rely on technology to guide the plane.

AIR POLLUTION

Cough, hack, wheeze! Where do you go for a breath of fresh air when you are surrounded by pollutants? Air pollution is any chemical in the air that can cause harm to people or other living things. Some common pollutants are smoke, carbon monoxide, nitrogen dioxide, sulfur dioxide, ozone, and lead. Many cause direct harm when animals breathe them or take them in through their skin. Others mix with harmless chemicals in the atmosphere to form acid rain or smog. Even chemicals that are not normally poisonous, such as carbon dioxide, can cause far-reaching environmental problems when given off in large amounts. People are working to reduce air pollution by using air filters and smokestack scrubbers in factories and power plants, and catalytic converters in cars. Alternative energy sources, environmentally friendly materials, and new production and disposal processes are also being developed. Governments are setting limits and charging fines to companies that produce pollution. International treaties, such as the Kyoto Protocol, organize the efforts of many countries together to reduce these harmful gases.

did you know?

A PERIOD OF EXTREME AIR POLLUTION IN LONDON, CALLED *THE GREAT SMOG OF 1952*, KILLED CLOSE TO 4,000 PEOPLE IN JUST 4 DAYS.

Wind can carry air pollution hundreds of miles and affect communities far from its source.

Steel manufacturing is one source of air pollution in Volta Redonda, Brazil.

WHERE DOES IT COME FROM? ▼

Pollutants can come from natural and human sources. Smoke is produced during a forest fire. Volcanoes produce sulfur dioxide and carbon dioxide. Carbon monoxide, nitrogen oxides, and sulfur dioxide come from car exhaust and gases released from burning fuel in power plants. Lead can come from industrial wastes and cars, and ozone is created when other pollutants react together in the atmosphere.

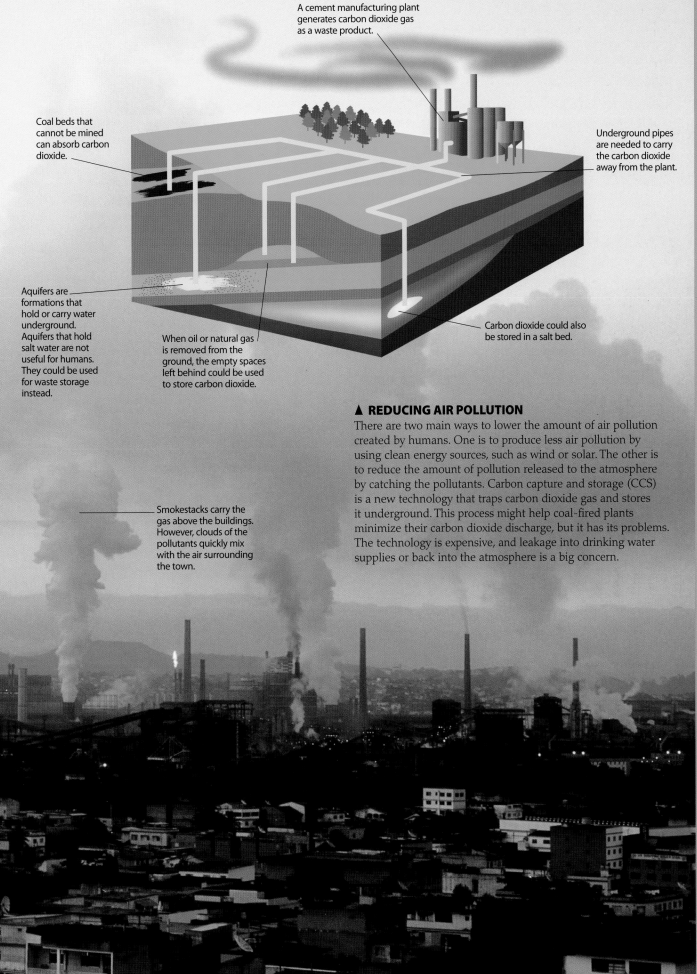

A cement manufacturing plant generates carbon dioxide gas as a waste product.

Coal beds that cannot be mined can absorb carbon dioxide.

Underground pipes are needed to carry the carbon dioxide away from the plant.

Aquifers are formations that hold or carry water underground. Aquifers that hold salt water are not useful for humans. They could be used for waste storage instead.

When oil or natural gas is removed from the ground, the empty spaces left behind could be used to store carbon dioxide.

Carbon dioxide could also be stored in a salt bed.

Smokestacks carry the gas above the buildings. However, clouds of the pollutants quickly mix with the air surrounding the town.

▲ REDUCING AIR POLLUTION

There are two main ways to lower the amount of air pollution created by humans. One is to produce less air pollution by using clean energy sources, such as wind or solar. The other is to reduce the amount of pollution released to the atmosphere by catching the pollutants. Carbon capture and storage (CCS) is a new technology that traps carbon dioxide gas and stores it underground. This process might help coal-fired plants minimize their carbon dioxide discharge, but it has its problems. The technology is expensive, and leakage into drinking water supplies or back into the atmosphere is a big concern.

ACID RAIN

It can cause the paint to rub off your car. It can eat away stone buildings and sculptures, poison trees, and even kill entire lakes full of fish. You might think that only very concentrated acids can cause this kind of harm. However, given enough time, acid rain can be quite a threat. When power plants, cars, and factories burn fuel, they emit gases such as sulfur dioxide and nitrogen oxide. Volcanoes, forest fires, and decaying plants emit these gases, too. When these gases enter the atmosphere, they react with harmless gases to form sulfuric and nitric acids. These acids combine with water vapor and fall back to Earth in the form of acid rain. Acid-rain-forming gases can travel with winds for hundreds of miles. That means that acid rain can have expensive and deadly effects both locally and far from where it formed.

Built around 420 B.C., this porch decorates an ancient Greek temple at the Acropolis of Athens, Greece.

WEARING AWAY STONE ▶

Rain and other kinds of weather slowly break down rocks over time. Acid rain can speed up this process. Normal rain can have an acidity, or pH, of 6. Acid rain, on the other hand, can be ten times more acidic—with a pH of 5. Acids react with stones such as marble, limestone, and granite to form softer materials that crumble away over time.

did you know? ....
THE UNITED STATES PRODUCES MORE SULFURIC ACID THAN ANY OTHER CHEMICAL—ABOUT 40 MILLION TONS.

ACID BURNS ▼

Concentrated sulfuric acid is an oily, colorless liquid. When the acid comes in contact with this paper, it reacts with carbohydrates, such as cellulose—a fiber that comes from wood and other plants. In much the same way that a hot fire burns wood, the reaction removes water molecules and leaves behind black, soot-like carbon.

Just as concentrated acid burns this paper, weaker acid rain slowly poisons and disintegrates living and nonliving things.

DESTROYING LIFE ▼

Acid rain that soaks into soil can dissolve and wash away nutrients that are important for plant growth. Trees weakened by acidic soil can lose their leaves. They can also become more susceptible to other environmental threats. Acid rain collects as runoff in rivers and lakes, killing fish and other aquatic organisms.

Column sculpted
from solid marble

Until it was moved to
a museum, acid rain
was dissolving this
ancient work of art.

GLOBAL WARMING

On a cold night, a blanket keeps you warm. In cold space, greenhouse gases like carbon dioxide (CO_2) and methane surround Earth and keep it warm. When the sun's rays enter the atmosphere, Earth's surface absorbs most of the heat; the rest radiates back into the atmosphere. Some of this radiated heat passes into space, but greenhouse gases trap most of it. Living things need a certain amount of this trapped warmth to survive. Burning fossil fuels—oil, coal, and natural gas—to power automobiles, factories, and homes releases substantial amounts of CO_2 into the atmosphere. Most scientists are confident that these CO_2 emissions increase the amount of greenhouse gas in the atmosphere, trapping too much thermal radiation. They have concluded that global warming—the increase in Earth's average surface temperature—leads to climate change that impacts life on Earth.

MELTING ICE ▼

Studies show that global warming is changing circulation patterns in the oceans and atmosphere. These changes, along with warming temperatures, contribute to the widespread melting and shrinking of glaciers. Scientists use satellite images and computer models to observe and predict changes in the rate of melting. Evidence indicates that glacial melting is accelerating. Melting arctic ice reduces the habitat of wildlife, such as polar bears.

did you know?......................

MORE THAN HALF OF ALL FOSSIL FUELS EVER USED HAVE BEEN CONSUMED IN JUST THE LAST 20 YEARS.

RISING SEA LEVELS ►

Global warming is causing sea levels to rise faster, partly because of the rapidly melting glaciers. At the same time, warmer water temperatures increase the volume of ocean water. This process is called *thermal expansion*. Average sea levels are expected to rise by 7 to 23 inches (about 18 to 58 cm) or more by the end of this century. Barrier islands and coastal wetlands may be lost, and coastal communities are at greater risk of flooding. The streets of Venice, Italy—a city historically prone to flooding—have some degree of flooding 200 days per year. If sea levels rise, that number could rise.

KEY
Land submerged if sea level rises

13-FOOT (ABOUT 4-M) RISE IN SEA LEVEL

Rising sea levels in Florida would affect cities and ecosystems.

26-FOOT (ABOUT 8-M) RISE IN SEA LEVEL

Much of southern Florida, including Miami, would be submerged if sea levels rose 26 feet (about 8 m).

Researchers are developing gates that will close Venice's three inlets against the flooding tides of the Adriatic Sea.

◄ BLACK CARBON POLLUTION

Incomplete combustion, or partial burning, of fossil fuels, biofuels, and biomass, such as wood, releases black carbon into the atmosphere. Black carbon is a type of tiny floating particle called an *aerosol*. Black carbon absorbs incoming solar radiation and contributes to atmospheric warming. Researchers estimate that, in the past 30 years, aerosols have caused 45 percent of the warming in the Arctic region.

EQUATOR

The equator is like a belt that circles Earth's middle. It is an imaginary line located equal distances from the North and South poles, at what is known as 0 degrees latitude. The equator is the longest line of latitude on Earth, measuring about 24,901 miles (about 40,075 km)! It is often used as a point of reference. The distance of each location from the equator determines how many hours of sunlight it receives each day and how direct that sunlight is. The areas near the equator are very, very hot! But, since other factors also affect weather, such as nearness to the ocean, mountains, height above sea level, and atmospheric conditions, temperature and rainfall in each place can vary a lot. That is why you can find both tropical rain forests and deserts near the equator.

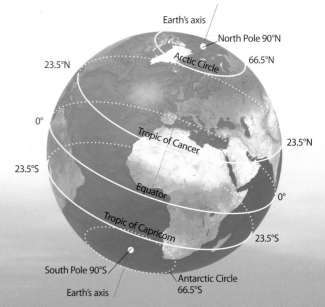

LINES OF LATITUDE ▲

Earth's axis is tilted slightly, relative to the sun. Places farther from the equator receive varying amounts of sunlight during the year, depending on which of Earth's hemispheres is facing the sun. The area near the equator, called the *tropics*, gets about 12 hours of sunlight each day year-round. The tropics lie between two other imaginary lines of latitude, the Tropic of Cancer and the Tropic of Capricorn.

Drought has lowered the level of Lake Turkana, concentrating the water's salt and other chemicals so the water is now barely drinkable.

AN AFRICAN DESERT

Many areas along the equator are parched and dry. Rainfall in this part of Kenya is typically less than 8 inches (about 200 mm) annually. Running through this desert landscape is a valley, called a *rift*, which formed as pieces of Earth's crust pulled—and continue to pull—apart. Lakes formed in these valleys. This village is on Lake Turkana, which is a large but shallow lake in the middle of a desert.

TROPICAL RAIN FOREST OF HAWAII ▶

The Hawaiian Islands lie just south of the Tropic of Cancer. The weather can vary dramatically within short distances. Honolulu can receive more than 20 inches (508 mm) of rain in a single year. Three miles inland, here in Manoa Valley, more than 150 inches (3,810 mm) fall. Like other tropical rain forests near the equator, this one is rainy, warm, and teeming with life.

did you know?..
TANZANIA'S MOUNT KILIMANJARO IS ONLY 3 DEGREES OF LATITUDE SOUTH OF THE EQUATOR—ONLY 207 MILES (ABOUT 333 KM)—BUT ITS PEAK IS CAPPED WITH SNOW.

Acacias, or umbrella thorn trees, provide shade, food, fuel, and more for the people who live near the lake.

DUNES

In a village in the West African nation of Mauritania, homes are in danger. Houses in this arid desert are being swallowed by advancing sand dunes. A dune is a landform in motion, constantly shifting position as wind erosion moves particles of sand from one location and deposits them in another. Dunes form when wind carries sand-size particles in a jumping or bouncing motion, a process called *saltation*. As the wind carries the sand, the wind slows down. The wind then deposits the sand, often forming ripples that correspond to the lengths of the jumps. The wind continues to pick up and deposit sand in this way. Larger particles can creep along in the desert when they are hit by these saltating particles.

▼ DUNE DIRECTION

Which way are these dunes moving? A dune's shape tells its story. Dunes move toward their leeward side—the side opposite from the wind direction. Generally, the windward slope is less steep than the leeward slope. Surface ripples are also revealing. Those on the windward side are longer and shallower, while the ripples on the leeward sides are shorter and steeper. In this photo, the steeper slopes are to the right, so the dunes are moving in that direction.

Wind pushes sand up the windward slope. To move sand, wind must be fast enough to overcome surface friction.

Wind deposits sand down the leeward side, called the *slip face*. Its steepness is determined by the size and shape of sand grains.

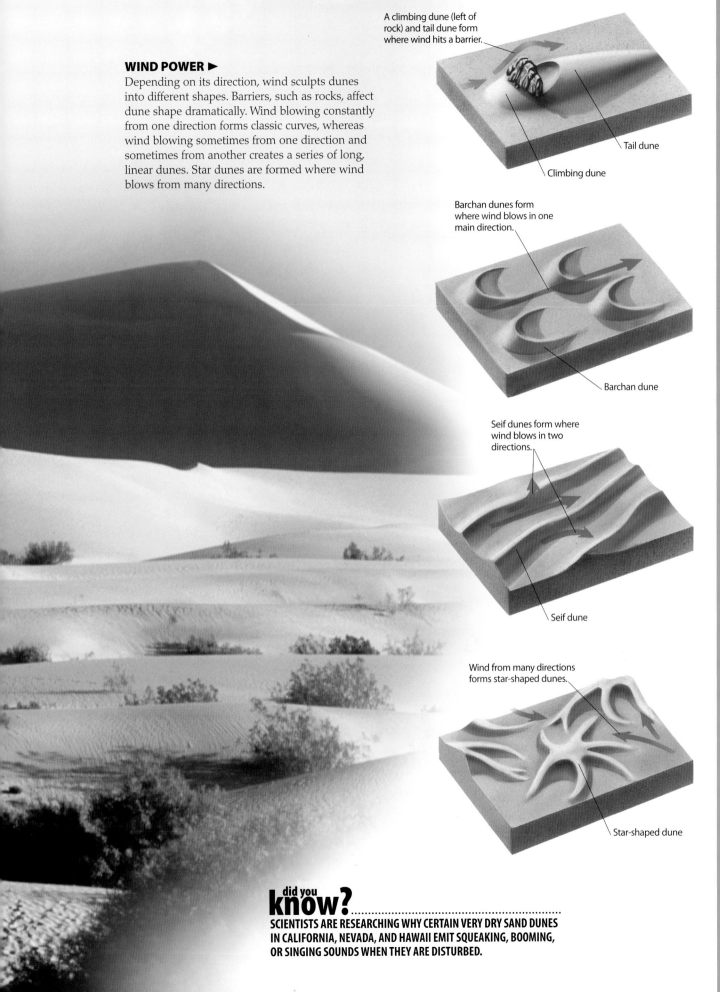

WIND POWER ▶

Depending on its direction, wind sculpts dunes into different shapes. Barriers, such as rocks, affect dune shape dramatically. Wind blowing constantly from one direction forms classic curves, whereas wind blowing sometimes from one direction and sometimes from another creates a series of long, linear dunes. Star dunes are formed where wind blows from many directions.

A climbing dune (left of rock) and tail dune form where wind hits a barrier.

Tail dune

Climbing dune

Barchan dunes form where wind blows in one main direction.

Barchan dune

Seif dunes form where wind blows in two directions.

Seif dune

Wind from many directions forms star-shaped dunes.

Star-shaped dune

did you know?

SCIENTISTS ARE RESEARCHING WHY CERTAIN VERY DRY SAND DUNES IN CALIFORNIA, NEVADA, AND HAWAII EMIT SQUEAKING, BOOMING, OR SINGING SOUNDS WHEN THEY ARE DISTURBED.

MIRAGES

In just about any cartoon that takes place in the desert, a hot, thirsty character sees an inviting pool of water in the middle of a burning expanse of sand. But what the character actually sees is a mirage, not a pool of water. A mirage is an image of a distant object that is caused by refraction—light waves changing speed as they pass from one medium into another, causing the light to bend. The mirage is real in the sense that it is an optical phenomenon, like a rainbow, that can be photographed by a camera. When you look out at an expanse of hot sand or highway, you might see a shiny, bluish surface on the ground. You are actually seeing an image of the sky, which looks as if there were a mirror lying on the ground. But the sky on the ground makes no sense, so your brain interprets the image as something that might actually form on the ground—a pool of water.

A LAKE IN THE DESERT?

Mirages are especially convincing in places like the Namib Desert in Africa. Mirages occur where the eye can see across a relatively flat surface for some distance. Here, the trees appear to be reflected in a lake. What looks like a lake is the result of light bending as it passes through layers of air that are at different temperatures. For a mirage to occur, the light has to be hitting the hot desert surface and the trees at a certain angle, typically from lower in the sky. The light that bounces off distant objects is refracted as it passes from the cooler air at eye level to the hotter air on the desert floor.

The mirage of the trees and sand is seen below the trees and sand, so it is called an *inferior mirage*.

Light refracted as it passes through the hot air above the desert appears as a lake.

WET ROAD AHEAD ▶

Where did that puddle come from on a hot, dry day? Black pavement doesn't reflect images, so it has to be water that you see, reflecting the motorcycles and car up ahead. As you draw closer, however, the water disappears. That's because there never was any water—you saw a mirage!

1. Light rays usually travel in a straight line.

2. This ray coming from the top of the tree begins bending when it hits the hot air.

3. The light ray is bent upward by hot air.

4. The bent light ray enters the eye in a straight line, so it appears to have originated from a point on the ground, as if it had traveled in a straight (dashed) line.

Real object

Layer of cool air

Layer of hot air

Mirage

did you know?.......................

MIRAGES CAN OCCUR ABOVE COLD SURFACES, SUCH AS THE OCEAN, ICE, OR SNOW. ON THE OCEAN, ISLANDS CAN APPEAR TO BE FLOATING ABOVE THE HORIZON. WHEN A MIRAGE IS SEEN ABOVE THE ACTUAL OBJECT, IT IS CALLED A *SUPERIOR MIRAGE*.

HOW IT WORKS ▲

Light rays bounce off a tree in the distance. The rays travel through layers of air. Usually the light rays travel in a straight line. The layer of air right next to the sand is much hotter than the layer of air above it. Light travels faster through hot air, which is less dense than cool air. The light rays that pass from cool air to hot air travel faster, so they refract. These rays bend upward toward the viewer's eye, rather than continuing on the path they were on in the cool layer, forming a mirage.

Light travels directly to your eye to form an image of the tree in the distance.

Light refracted through the layer of hot air creates an inverted image.

ATACAMA DESERT

The driest place on Earth is right next to the Pacific Ocean. It's the Atacama Desert, a narrow plateau that stretches more than 620 miles (1,000 km) along the coast of Chile. In some parts of the Atacama, rain has never been recorded! This is because the desert is bordered on the east by the high Andes Mountains and on the west by a coastal mountain range. The Andes block warm, wet air from reaching the desert. The air rises when it hits the mountains, cools, and then falls as snow in the mountains and rain in the Amazon rain forest. By the time the air reaches the Atacama Desert, it is dry. The air that blows in from the cold Pacific is dry, too.

SPINY SURVIVORS ▶

Although the Atacama's climate is extremely dry, a few plants have adapted to desert life. Many are unique to the region. Cactuses are some of the few organisms that can live in the Atacama's harsh conditions. Their stems are covered in a waxy coating that keeps the plant from losing too much water through its pores. Cactus roots are also close to the soil surface, so they can quickly suck up rain when it finally falls.

◀ **A MOONSCAPE**

Because it rarely rains in the Atacama, the soil is not moist enough to support much plant life. Few animals can survive here because there is little to eat. In fact, in the Atacama it rains on average only .004 inches (.01 cm) a year. By comparison, New Orleans receives an average of 58 inches (147 cm) of rain per year. Many parts of the Atacama are so barren, or lifeless, that they are often compared to the surface of the moon or Mars. NASA scientists found so little life in Atacama's soil that they suggested it would be a good place to test the equipment they use when they search for life on Mars.

This candelabra cactus grows less than 1/4 inch (5 mm) annually, about the length of an eyelash!

The air in the Atacama is so dry that clouds here are rare. The most significant source of moisture to the Atacama Desert is fog from the Pacific Ocean.

Unlike hot deserts such as the Sahara, the Atacama is relatively cool. Average daily temperatures range from 32°F to 77°F (0°C–25°C).

Human remains that are 9,000 years old have been found in the Atacama Desert. The dry air slows decomposition and keeps the mummies extremely well preserved.

did you know?......................
THE DRIEST PLACE ON EARTH IS JUST WEST OF ONE OF THE WETTEST PLACES ON EARTH—THE AMAZON BASIN.

AMAZON RIVER

The Amazon River practically cuts South America in half. It starts in the Andes Mountains, close to the Pacific Ocean, and flows almost 4,000 miles (6,400 km) across the continent to the Atlantic. For most of its length, the Amazon pushes lazily through the gentle slopes of the forest. Don't let that pace fool you, though. The Amazon carries more water than any river in the world. In some places it is so wide that you cannot see from one bank to the other. It carries trillions of gallons of water—more than ten times the flow of the Mississippi River. Even so, it has incredible force. When it reaches the Atlantic Ocean, the Amazon River's fresh water pushes the ocean's salt water for more than 100 miles (161 km) before they mix.

DRAINING A BASIN ▼

The Amazon River, and the rivers that feed into it, drain a vast area of South America. This area, called the Amazon basin, covers more than 2.7 million square miles (7 million sq km)—almost as much as the continental United States. The Amazon basin is home to thousands of species of animals. In the river, manatees munch on aquatic plants. Freshwater dolphins use the echoes of their high-pitched voices to find fish. Along the riverbank, you might find large hunters, such as the jaguar, the most powerful cat in the Americas.

One river cannot drain the basin by itself. More than 1,000 other rivers dump their waters into the Amazon, including the one shown here, the Tigre.

The Amazon is a river of many colors, from black to muddy yellow, depending on each region's soil and plants.

About 20 percent of the world's supply of fresh water flows down the Amazon River to the Atlantic Ocean.

THE TOP OF THE RIVER ▶

The waters of the Amazon and its tributaries aren't all calm and quiet. The river begins in the high mountains of Peru. As gravity pulls the water, it tumbles downward with amazing force, scouring rocks and carving deep canyons. The water pours over cliffs, creating crashing waterfalls, some of them hundreds of feet high. The sediments from the high river nourish the land far downstream. Don't try rafting here!

did you know?....................
ALTHOUGH THE AMAZON CUTS ALL THE WAY ACROSS BRAZIL, NOT A SINGLE BRIDGE CROSSES THE RIVER.

Between 60 and 140 inches (about 1,500–3,500 mm) of rain fall each year in the Amazon basin.

There are more than 2,000 species of fish in the Amazon River, greater than the number of species of fish in the entire Atlantic Ocean.

Although most of the more than 20 species of piranha in the Amazon are vegetarians, it is the red-bellied piranha, with its sharp teeth, that stars in jungle movies.

The rain forest creates its own rain. About half of the water vapor in the air comes from the leaves of plants.

RAIN FOREST

When you imagine a rain forest biome, do you picture a hot, humid jungle filled with colorful flowers and birds; noisy insects; and spectacular snakes, frogs, and mammals? You might think rain forests exist only in the tropics, near the equator—but they are on every continent except Antarctica. Our own Pacific coast, from northern California into Canada and Alaska, is home to the largest temperate rain forests. Rain forests are defined by temperature and rainfall. Tropical rain forests receive between 72 and 360 inches (about 183–914 cm) of rain a year and have temperatures of about 80°F (about 27°C) and higher. Temperate rain forests are cooler and average 60 to 200 inches (about 152–508 cm) of rain a year. The differences in rainfall and temperature create two distinctive rain forest biomes, each with its own structure and wildlife.

did you know? ABOUT 70 PERCENT OF PLANTS THAT HELP FIGHT CANCER COME FROM THE TROPICAL RAIN FOREST.

◄ **GOLDEN LION TAMARIN**
Weighing from 14 to 29 ounces (about 396–822 g), this species of small monkey lives in family groups in the canopy of Brazilian rain forests. For food, they depend on organisms and water inside certain plants, as well as fruit, insects, and small lizards. The golden lion tamarin is critically endangered because its habitat has been destroyed by deforestation.

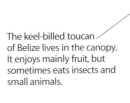

The keel-billed toucan of Belize lives in the canopy. It enjoys mainly fruit, but sometimes eats insects and small animals.

◄ GIANTS OF THE JUNGLE

Both tropical and temperate rain forests are divided into layers. Rain forests have an emergent layer, in which the tallest trees poke through the canopy. Growing to more than 200 feet (about 61 m), the trees of the tropical rain forest have ample space around them to spread their foliage. Between intense rainstorms, the emergent layer is exposed to extreme, drying sunlight and winds. The emergent trees of the temperate rain forest can grow to heights greater than 328 feet (100 m) due to their long lives—they usually live over 500 years!

ADAPTABLE ANIMALS ►

Tropical and temperate rain forest animals compete for food and shelter, so they must adapt to survive. Monkeys developed prehensile tails that can grasp branches. Other animals, such as this tree frog, sport bright colors to warn off predators or patterns to blend in with the forest. Still others have adapted a specialized diet and eat only one or two foods.

◄ THE FOREST ROOF

The dense branches and foliage of the tropical rain forest canopy shield the layers below from direct sun, wind, and heavy rain. Insects, birds, monkeys, snakes, and frogs feast on the abundance of fruits, flowers, leaves, seeds, and nectar. Fewer creatures live in the redwoods and other tall evergreen conifers—trees that produce cones and seeds—that dominate the temperate rain forest canopy.

Growing to more than 5 inches (13 cm) in length, the white-lipped tree frog of Australia is the largest tree frog in the world.

199

RAIN FOREST CONTINUED

Rain forests cover about 2 percent of Earth's surface, yet more than 50 percent of all plant and animal species live in them. Rain forests are a valuable, yet fragile, resource. They regulate global temperatures and weather patterns and help maintain Earth's limited supply of fresh water. Native cultures that live in tropical rain forests depend on resources from this environment for their survival. Valuable products, such as timber and coffee, are important exports, but they come at a price. Local and global companies with farming, timber, and ranching interests are deforesting the rain forest at an alarming rate. Although some deforestation is necessary to build homes and create agricultural areas, practices like clear-cutting and burning vast areas of land, as well as mining, are destructive. Rain forest destruction threatens biodiversity, promotes flooding, and causes soil erosion.

THE DARK MAZE ▶

Very little sunshine penetrates through to the understory layer of the tropical and temperate rain forest. Shade-loving plants with large leaves are home to insects such as beetles, bees, and ants. Snakes, lizards, and spiders can hide in the dense overgrown shrubbery.

JAGUAR ▼

Jaguars are fierce predators in the tropical rain forests of Central and South America. They hunt mainly on the ground, but sometimes pounce from tree limbs onto unsuspecting prey such as crocodiles, snakes, monkeys, and large piglike animals called *tapirs*. Jaguars are considered "near-threatened" because of deforestation and poaching.

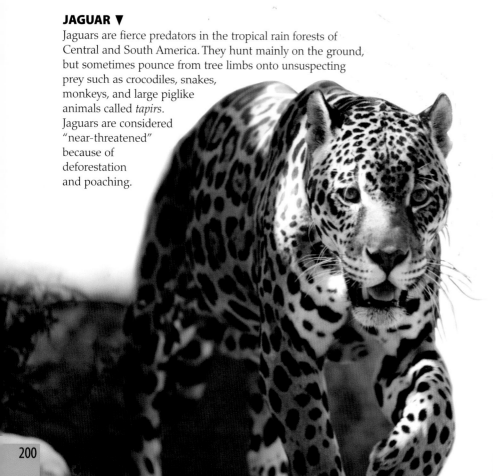

The northern spotted owl is considered a threatened species. Its habitat has been severely reduced by logging of temperate rain forests in the United States and Canada.

Leafcutter ants bite off and carry away bits of leaves. They use the leaf bits to grow a fungus that they eat

LIFE IN THE UNDERSTORY ▶

Flowers in the tropical understory are large, pale, and heavily scented to attract pollinating insects. Many grow on tree trunks, unlike temperate rain forest flowers. Camouflaged reptiles and insects can easily hide from predators in the dim light.

◀ A SOFT LANDING

The soil of the temperate rain forest is rich and moist, covered with decaying needles and leaves. Wildflowers, grasses, mosses, and toadstools grow here. On the crowded floor of the Alaskan rain forest shown here, seeds can fall onto dead trees. Seedlings can take root right on the decaying trees, which are called *nurse logs*. The tropical rain forest floor is dark and the air is almost still. Little more than fungi and plants that get their nutrients from decaying leaves can grow in such darkness.

CREATURES OF THE FOREST FLOOR ▶

Termites, scorpions, and other invertebrates live on the tropical rain forest floor. Foragers hunt bugs and edible roots. Most temperate rain forest creatures live on or near the floor, where most of the food is. Chipmunks, birds, deer, black bears, and cougars are among these animals.

Its pattern helps the tropical Gaboon viper hide on the forest floor, waiting for its next meal.

did you know?
GABOON VIPERS HAVE THE LARGEST FANGS OF ALL SNAKES. THE FANGS CAN GROW AS LONG AS 2 INCHES (ABOUT 5 CM).

MOUNT EVEREST

The top of Mount Everest is the highest place on Earth. Its snowcapped peak stands at 29,035 feet (almost 8,850 m) above sea level—and it is still growing! Colliding tectonic plates are pushing the mighty Himalaya Mountains up at a rate of about 5 millimeters per year. Weather conditions at the summit can be extreme, with hurricane-force winds and an average temperature of –33°F (–36°C). Most climbers try for the summit in the spring when conditions are most favorable. But even then, fierce storms can happen suddenly. Even in good weather, glaciers can shift and crack, creating dangerous crevasses. With only one third of the oxygen at sea level available to them, many climbers have died because of these treacherous conditions.

The North Face is one of Everest's three faces, or sides.

Rongbuk glacier

THE YAK ▲
Yaks are huge shaggy beasts well suited to life in the mountains. They have adaptations, such as large lungs and more red blood cells, that allow them to live at high altitudes where oxygen is scarce. Yaks are important to the survival of the people of the Himalayas. They are a source of milk, meat, and fur for warm coats. Their dung is dried and used as fuel, and they are excellent mountain pack animals.

SNOW LEOPARDS ▶
Snow leopards are an endangered species and are very rarely seen in the wild. Between 3,000 and 6,000 of these amazing cats are left in the vastness of the Himalayas. They have been hunted for their beautiful thick fur, and their bones and organs are used in traditional Chinese medicine. Only one snow leopard has been seen on Mount Everest in the past 40 years.

The South Face is the most popular route to the summit.

The border between Nepal and China runs across the top of Everest.

A MOUNTAIN WITH MANY NAMES

In 1865, Mount Everest was named after Sir George Everest, the British Surveyor General of India at the time the mountain was first mapped in 1841. In Nepalese, it is called *Sagarmatha*, "forehead in the sky," and in Tibetan, *Chomolungma*, "goddess mother of the world." The summit was first reached in 1953 after many failed attempts. Now, there are many expeditions up the mountain each year—but not all of them are successful.

▲ A BREATH OF THIN AIR

Climbing Mount Everest is no easy task. Climbers must spend several weeks adapting to the high elevation at Base Camp—around 17,500 feet (5,334 m). They then rest at higher camps over 21,000 feet (6,400 m), acclimating to the lower oxygen levels as they ascend. Most climbers need extra oxygen to reach the summit. Lack of oxygen flow to the brain can make them feel dizzy, sleepy, and confused. At high altitude, this is a serious risk to their lives.

did you know? THE ROCKS THAT FORM THE HIMALAYAS, INCLUDING MOUNT EVEREST, WERE ONCE PART OF THE OCEAN FLOOR!

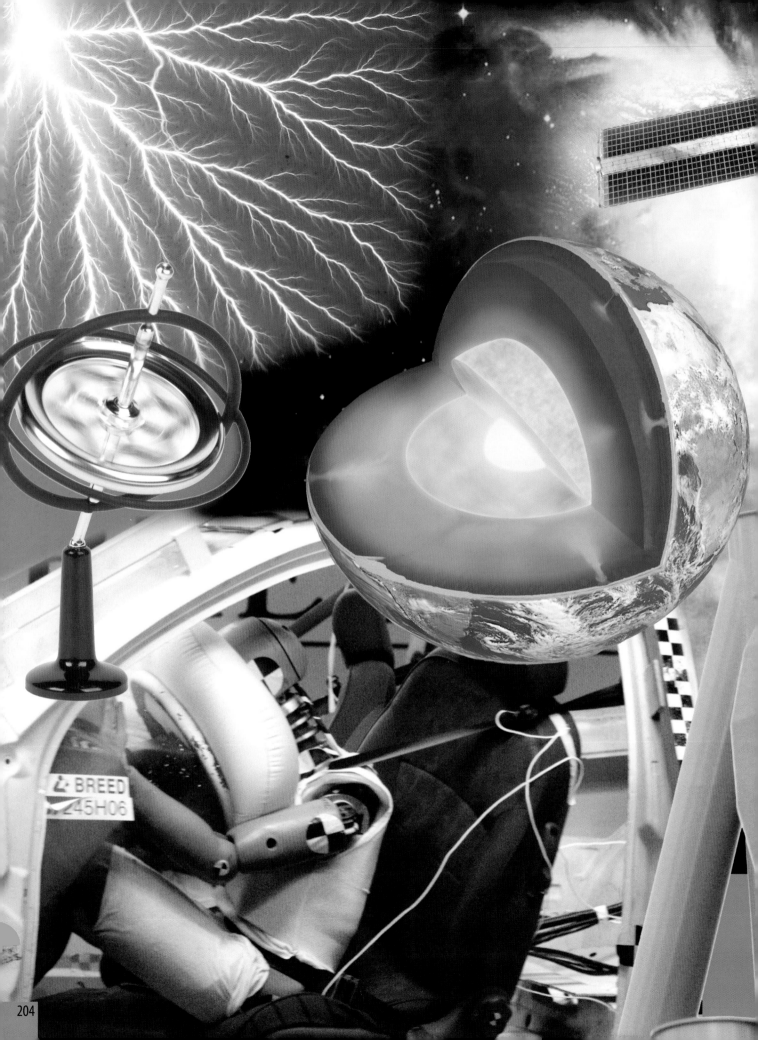

PHYSICAL SCIENCES

From the tiny atom, to the depths of the universe, physical science studies how and why non-living things behave the way they do. With universal laws, it explains processes that we cannot see, but are constantly working in our world, for example, the electricity powering our homes, the sound of an airplane, or the forces that move a roller coaster car. It looks at the intricate structures of materials, and how they interact. So scientists can manipulate atoms to make medicine or fireproof clothing. Beyond Earth, physical science examines the matter in the universe, including stars, planets, galaxies, and all that lies between. A space so vast, that by the time light from distant objects reaches our telescopes, we are really looking at a view from back in time.

GRAVITY

No matter where you stand on Earth's surface, gravity pulls you toward the center of the planet. As you stand, the ground is pushing upward on your feet in reaction to the gravitational force. In deep space, where the gravitational pull of stars and planets is very small, astronauts experience a sense of weightlessness, a condition called *microgravity*. It's possible to feel this sensation on Earth, if only for a short time during a jump off the ground. You can feel reduced gravity by taking a fast elevator ride down in a tall building. If you jump inside the elevator, it takes longer than usual to come to the floor because the elevator is falling, too.

Space scientists study the growth of substances, such as these salt crystals, in microgravity to learn more about the properties of materials.

SCIENCE IN SPACE ▼

The International Space Station is constantly being pulled toward Earth by gravity. Everything inside the station is being pulled at the same rate. This creates a condition of microgravity that is ideal for scientific experiments. In Earth's gravity, less dense, hot gases and soot from a candle flame rise. The soot gives the flame its yellow color. Cooler, denser gases sink, bringing oxygen to the base of the flame. In microgravity, the flame spreads in every direction, because there is no force moving the gases. Very little oxygen moves to the flame, so the flame's temperature drops. The cooler flame produces very little soot, so it burns blue.

THE FLOATING SCIENTIST ▲

Astronauts inside the space station are being pulled by gravity at exactly the same rate as the station itself, so they don't fall to the floor. They train for orbit in large jet airplanes that fly up and down in curving paths. The plane, known as the Vomit Comet, and its contents are falling at the same rate for about 25 seconds, simulating microgravity. Physicist Stephen Hawking, floating above, experienced microgravity during this type of flight.

▲ COMING DOWN TO EARTH

As the motocross rider jumps from the ramp, his muscles provide a force that moves him upward. Once he is in the air, the force of gravity takes over and he falls. During the jump, the boy and the bike fall at the same rate, so he experiences a brief feeling of weightlessness. During that time, he can perform a trick called a *cordova* while doing a back flip, and land—all in a fraction of a second.

**did you
know?**.....................................
WANT TO BE TALLER? SPEND TIME IN MICROGRAVITY. WHEN THERE IS NO DOWNWARD PRESSURE ON THE SPINE, ASTRONAUTS BECOME ALMOST 2 INCHES (5 CM) TALLER.

BRIDGES

To cross over a creek, you find a log that reaches from one bank to the other.
The log holds up your weight and keeps your feet out of the water—as long as
you can balance! You are demonstrating Newton's third law of motion: for every
action, there is an equal and opposite reaction. Your body pushes down on the log.
The log and the points on the banks where it rests react by pushing up with equal
force. The weight the bridge can hold is called the *load*. The length of the bridge is
the *span*. Throughout history, people have found ingenious ways to carry heavier
loads over longer spans. Both bridges shown here use one of the oldest forms, an
arch, to help support the load across the span.

BRIDGES AND HAMMOCKS ▶

Think of a hammock, held up by ropes attached to trees. This
suspension bridge is held up by two giant cables—3 feet (about 1 m)
in diameter. The cables are draped over towers and attached on each
end to a 60,000-ton concrete structure, called an *anchorage*, where
the bridge is attached to the ground. Smaller cables suspend the
weight from the two main cables. The cables distribute the load over
the towers and out to the anchorages.

From end to end, the
Golden Gate Bridge
stretches 8,981 feet
(2,737 m) across San
Francisco Bay.

The two main cables
pull like a bow string
on the anchorages.

The main towers reach 746 feet (227 m) above the water and about 110 feet (34 m) below the water.

Here, the force pushing down on the long arms of the cantilevers is balanced by the weight of the structures at each end of this span.

CANTILEVERS AT WORK ▲

The Firth of Forth Bridge in Scotland, at 8,276 feet (2,523 m) long, is one of the world's longest cantilever bridges. A cantilever is a type of lever whose long end sticks out over the water, like a huge, very stiff diving board. The short end is held down by enormous weights. Two cantilevers, one coming from each end of a span, can be joined in the middle to form a bridge.

The center span of the bridge, between the towers, is the suspended portion. At 4,200 feet (1,280 m) long, it is now dwarfed by Japan's Akashi-Kaikyo bridge—the longest suspension—at 6,532 feet (1,991 m).

did you know?
ON THE GOLDEN GATE BRIDGE, THE MAIN CABLES SIT ON THE TOP OF THE TOWER IN HUGE STEEL SADDLES.

GRAVITRON

The carnival ride starts to spin faster and you feel yourself pressing against the wall behind you. You seem to be getting heavier. It's hard to pull your arms away from the padded wall. Suddenly, the floor drops away! You are sure you will fall, but you just keep spinning. The Gravitron is a popular carnival ride that uses a force known as *centripetal force* to give you a thrill. An object keeps moving in a straight line unless a net force acts on it. When you are whirling in a Gravitron, you have a net force acting on you that causes you to move in a circular path rather than flinging you in a straight line. That force is centripetal force. This same force keeps the moon and satellites in near-circular, or elliptical, orbits. In this case, gravity is pulling these bodies toward Earth's center. In the Gravitron, the wall you are pressed against allows centripetal force to keep you in your orbit!

RIDING THE GRAVITRON ▶

In a Gravitron ride, you feel as if you are being pushed against the wall, but the force is really just the opposite. If the wall were not there, you and your fellow passengers would fly off the ride the way a baseball flies out of a pitcher's hand.

The momentum of the moving disk keeps the gyroscope at the same angle. It takes a lot of force to change the motion.

GYROSCOPE ▲

A *gyroscope* is a device with a disk that spins rapidly about its axis. Gyroscopes are useful for studying the forces of circular motion, and they have important applications in air, sea, and land navigation. They are also used in the International Space Station and the Space Shuttle to maintain correct orientations.

did you know? RIDERS ON THE GRAVITRON MOVE AROUND IN A CIRCLE AT ABOUT 40 MILES PER HOUR (ABOUT 64 KM/H).

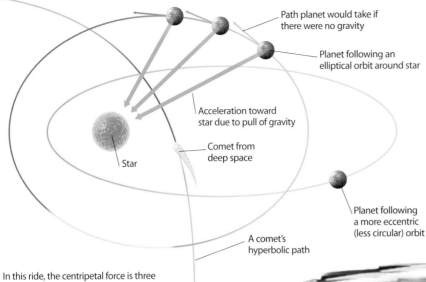

Path planet would take if there were no gravity

Planet following an elliptical orbit around star

Acceleration toward star due to pull of gravity

Comet from deep space

Star

Planet following a more eccentric (less circular) orbit

A comet's hyperbolic path

In this ride, the centripetal force is three times the force of gravity. Astronauts train in similar rides that have a centripetal force almost seven times as strong.

◄ HOLDING ONTO THE PLANETS

Planets moving around a star are held in place by centripetal force. That force is created by gravity. Without gravity, the planet's inertia, or tendency to keep moving in a straight line, would cause it to fly away into space. The mass of the star constantly pulls the planet inward. If inertia is greater than the pull toward the star, an object such as a comet does not orbit, but instead follows a hyperbolic path, which is elongated and not a closed curve. The comet eventually leaves the star far behind.

ROLLER COASTER

Clink, clink, clink. The roller coaster car slowly climbs its steep track. You grip the restraint in anticipation of a great fall. The thrills are about to begin! If you have ever been on a traditional roller coaster, you may have noticed that the cars do not have engines. They are pulled up to the highest point of the track by chains and then released. What causes them to go down and up again? Roller coasters can move along their tracks at great speeds because potential energy is converted to kinetic energy. Kinetic energy is the energy an object has due to its motion. Potential energy is energy related to an object's position or shape. In the case of a roller coaster, the potential energy is due to gravity and the height of the car. As the car falls, potential energy is converted to kinetic energy. The roller coaster car moves fastest at the bottom of a hill or a loop because all of the potential energy has changed to kinetic energy.

LOOP FORCES ▲

When a roller coaster car enters a loop, resistance to a change in motion, called *inertia*, keeps passengers in their seats. This continues as the car changes direction throughout the loop. When the passengers are completely upside down, the force of gravity acts on them, and they begin to fall. Luckily, the speed of the car keeps them moving in a circular path.

Roller coasters that go upside down may have a restraining bar that goes over the shoulders as well as the lap of the rider.

HEIGHT VERSUS SPEED ▲

At the highest point on a track, the roller coaster car has all of its potential energy and no kinetic energy. It is completely still. As the car falls, its potential energy gets converted to kinetic energy, causing its speed to increase. At the bottom of the hill, the roller coaster car is going as fast as it can. The car has no more potential energy because its potential energy has all been converted to kinetic energy. The kinetic energy is converted to potential energy again as the car's momentum carries it up the next hill. Each hill or loop must be lower than the one before it, because some of the original potential energy is lost due to air resistance and friction between the car and its track. Engineers design the different heights along the roller coaster track carefully so that a car does not get stuck due to a lack of energy.

Steel roller coasters have tracks that are usually pairs of metal tubes. Wheels on either side keep the car firmly attached to the track.

did you know?
THE FASTEST STEEL ROLLER COASTER IN THE WORLD, KINGDA KA IN JACKSON, NEW JERSEY, REACHES SPEEDS GREATER THAN 120 MILES (190 KM) PER HOUR.

COLLISION

The pitch and . . . the swing. There's a sudden, loud crack and the ball soars toward the fence. This is the kind of collision you hope for when you're playing baseball. Both the ball and the bat are moving, so they have what is called *momentum.* Momentum equals the mass of an object times its velocity. Velocity is speed with direction. During a hit, some of the bat's momentum is transferred to the ball. Since the mass of the ball does not change, this momentum goes into changing the ball's velocity. Instead of moving toward home plate, the ball is now soaring toward the right field fence. The total momentum of the bat and ball is conserved—that is, the total is the same before and after the collision. So the bat's momentum must decrease by the same amount as the ball's momentum increases by. So what's the secret to hitting a home run? Transfer enough momentum to hit the ball out of the park.

HEROIC DUMMIES ▶

A car crash is called an *inelastic collision.* When cars collide, they don't bounce off one another. Instead, metal crunches and bends, glass breaks, and sound waves carry energy away. Human bodies are not designed to absorb these sudden changes in energy and momentum. Crash test dummies are built to find out what happens to people in a car crash. Scientists use the information to design systems to protect passengers.

did you know? ....
MANY SCIENTISTS BELIEVE THAT THE MOON FORMED AFTER A COLLISION BETWEEN EARTH AND A MARS-SIZED ASTEROID BILLIONS OF YEARS AGO.

ELASTIC BUMPERS ▲

In what is called an *elastic collision,* the colliding objects bounce off of each other and don't change shape. All of the energy goes into changes in speed. Bumper boats are a great way to feel an elastic collision. As your boat smacks into your friend's boat, you both go spinning away from the collision.

Crash test dummies come in many sizes and shapes— adult men and women, children, infants, and even pregnant women.

Airbag

Joints are designed to move as human joints do.

The airbag fills with gas in a fraction of a second—faster than the blink of an eye.

As the airbag inflates, the seat belt locks in place, so that the dummy—or a person—does not keep moving forward as the car stops.

▲ FALLING INTO A BALLOON

If a car stops suddenly, the people in it will keep moving until something stops them. The soft surface of an airbag stops a person more slowly than either the dashboard or steering wheel. This results in less force on the person.

CATAPULTS

What do a kangaroo, a fishing lure, and a pole-vaulter have in common? They are all launched into the air with the help of a simple machine: the lever. A lever that is used to throw things is called a *catapult*. If you have ever put a marshmallow in a plastic spoon, bent the spoon back, and then let go, you've operated a catapult. Levers, like seesaws, lift things. If a lot of force is applied quickly to one arm, the lever not only lifts but also throws whatever is on the other arm. Most catapults have one long arm and one short one. They were used as ancient weapons to hurl large stones or other objects at enemies. A stone would be placed on the end of the long arm. When many people pulled a rope attached to the short arm, the long arm swung up, launching the stone up and at the enemy.

The pole is made of strong materials so that it springs back with great force when bent.

Australian Sophie Lichoudaris pole vaults during a 2009 athletic festival in Sydney.

POLE VAULT ▶

In the sport of pole vault, athletes use a long pole to launch themselves over a raised bar. Holding one end of the pole, the athlete runs toward the bar and places the far end of the pole in the ground just below the bar. The pole becomes a long lever that swings the athlete up and over the bar.

CASTING LONG ▼

A fishing rod uses the power of the catapult to cast a lure out into the water, and then uses the rod as a lever to haul in a fish. The angler's right hand acts as the balance point, called the *fulcrum*. With a quick motion, he pushes on the short arm of the catapult with his left hand. The rod pivots around his right hand, lifting the lure into the air and flinging it out across the water. If a fish bites, the rod becomes a lever that lifts the catch out of the water.

A hopping kangaroo can travel more than 35 miles (56 km) per hour.

The long, heavy tail helps the animal balance.

KANGAROO HOPS ▲

Kangaroos' legs use a lever action to launch them 30 feet (9 m) through the air. They have an especially long and springy Achilles tendon, which is a rubbery band that attaches leg muscles to heel bones. The tendon stores and releases energy so effectively that kangaroos get more bounce and use less energy when they speed up their hopping.

The raised bar is set to release easily, preventing injury should the athlete bump into it.

did you know?
POLEVAULTER SERGEI BUBKA CLEARED A 20-FOOT (6 M) BAR—ABOUT THE HEIGHT OF A TWO-STORY HOUSE WITHOUT THE ROOF.

Two vertical poles hold up the bar.

LIFTING ELECTROMAGNETS

Picking up your little sister's toy truck with a large magnet can make you feel very powerful. Imagine how exciting it would be to pick up a full-size car or a train. To do this, your tool would be an electromagnet. An electromagnet gets its strength from two sources. One source is a solenoid—coiled wire with a current running through it. The other source is ferromagnetic material— material that can become magnetized—such as iron, which forms a core inside the solenoid. The power of these two magnets working together can be hundreds—even thousands—of times greater than a magnet alone. Electromagnets can lift by way of the attraction of opposite poles. They can also lift by using the force of like poles repelling each other, allowing a train to float on air. Electromagnets can do more than lift. For example, they can lock and unlock a car door by turning an electric current on and off. They can move parts of an audio speaker, causing vibrations that produce sound waves.

LIFTING THE MAGLEV ▼

This train in Shanghai, China, is powered by magnetic levitation, called *maglev*. Maglev trains have a magnetic coil that runs along the track, forming what's called a guideway. When a current runs through the magnetic coil in the guideway, it repels a magnet along the bottom of the train, lifting the train about 0.39 to 3.94 inches (1 to 10 cm). Because the train is floating on air, it is not slowed down by friction with the ground.

The floor of the train, which has no wheels, glides smoothly because it is not in contact with the guideway.

The train wraps around each side of the guideway the way your hands wrap around a tray, so it can't fly off.

The train's electromagnets are on the part of the train that wraps under the guideway.

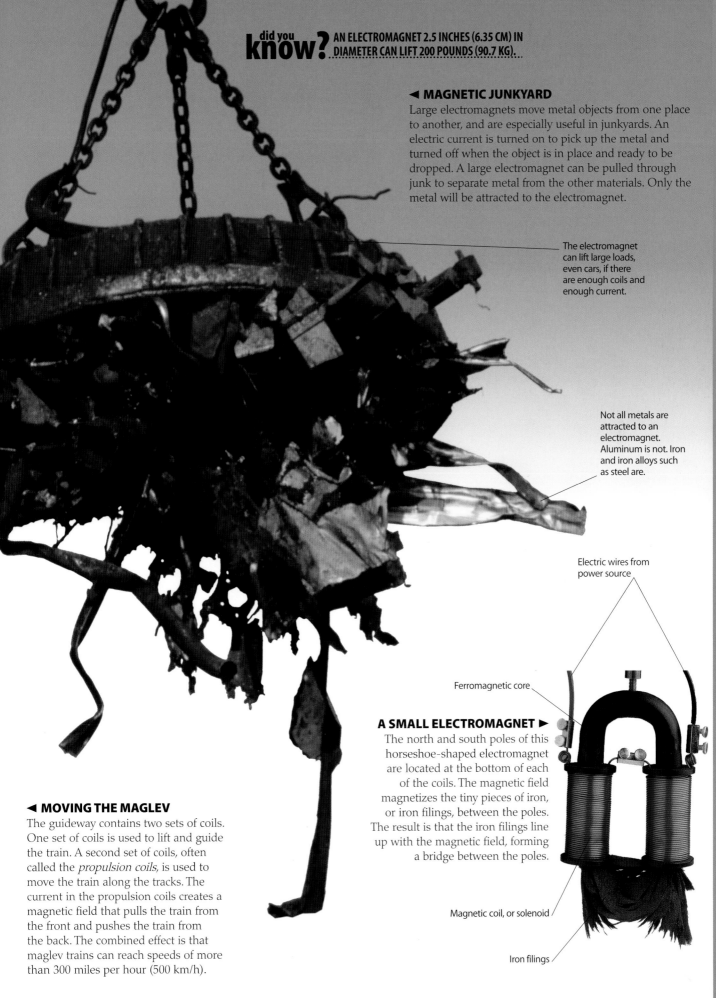

did you know? AN ELECTROMAGNET 2.5 INCHES (6.35 CM) IN DIAMETER CAN LIFT 200 POUNDS (90.7 KG).

◄ MAGNETIC JUNKYARD

Large electromagnets move metal objects from one place to another, and are especially useful in junkyards. An electric current is turned on to pick up the metal and turned off when the object is in place and ready to be dropped. A large electromagnet can be pulled through junk to separate metal from the other materials. Only the metal will be attracted to the electromagnet.

The electromagnet can lift large loads, even cars, if there are enough coils and enough current.

Not all metals are attracted to an electromagnet. Aluminum is not. Iron and iron alloys such as steel are.

Electric wires from power source

Ferromagnetic core

A SMALL ELECTROMAGNET ►

The north and south poles of this horseshoe-shaped electromagnet are located at the bottom of each of the coils. The magnetic field magnetizes the tiny pieces of iron, or iron filings, between the poles. The result is that the iron filings line up with the magnetic field, forming a bridge between the poles.

Magnetic coil, or solenoid

Iron filings

◄ MOVING THE MAGLEV

The guideway contains two sets of coils. One set of coils is used to lift and guide the train. A second set of coils, often called the *propulsion coils*, is used to move the train along the tracks. The current in the propulsion coils creates a magnetic field that pulls the train from the front and pushes the train from the back. The combined effect is that maglev trains can reach speeds of more than 300 miles per hour (500 km/h).

COLOR

Why is an orange orange and a blueberry blue? The things we see have a spectacular array of colors because of visible light. Light travels in waves. Light waves have different distances between the peaks of their waves. This distance is called a *wavelength.* You see different wavelengths as different colors. Together, all of these colors make up white light. But when light hits an object, some colors are absorbed and others are reflected. An object takes on the color it reflects. So we see a blueberry as blue because it reflects light with a blue wavelength.

VIOLET

Violet is the name given to the light with the shortest wavelength. *Purple* is the name given to the color you get when you mix blue and red pigments.

THE VISIBLE LIGHT SPECTRUM

The colors that the human eye can see make up what's called the *visible light spectrum.* This range includes every color in the rainbow. Red light has the longest wavelength. As you move counterclockwise around this color wheel from red, wavelengths decrease.

ORANGE AND YELLOW

Pigments are molecules that give things color by absorbing some colors and reflecting others. Carrots and bananas, for example, contain a pigment called carotene, which absorbs all colors of light except for orange and yellow.

did you know? THE PIGMENT CALLED *ANTHOCYANIN* MAKES EGGPLANTS LOOK PURPLE AND GIVES BLUEBERRIES THEIR COLOR.

DARK BLUE

The blue pigment known as *ultramarine* comes from blue semi-precious stones called *lapis lazuli.* Renaissance painters often saved this precious pigment to color only the most important people in their paintings.

IRIDESCENT BLUE

The blue morpho butterfly's color is not caused by a pigment. Tiny ridges on its wings cause the wings to reflect the blue wavelength of light. The wings seem to sparkle as the light bounces at different angles. This effect is called *iridescence.*

GREEN

The pigment that makes plants green is called *chlorophyll.* Chlorophyll absorbs all colors of light except green. When leaves change color in the fall, it is because the chlorophyll has begun to break down, and the other pigments in the leaves can be seen. These pigments reflect red, orange, and yellow light.

RADIO

In a thunderstorm, you see the lightning before you hear the thunder, because sound travels much more slowly than light does. How does music travel at the speed of light to your radio? By riding on radio waves! Like visible light, radio waves are electromagnetic waves that travel at the speed of light. Radio waves can travel long distances without being scattered or absorbed. Radio stations transform sound waves into radio waves. They modulate, or change, the radio waves so that they represent the sound of speech and music. Those waves are transmitted from a radio tower. Radios in homes and cars receive the waves and convert them back into sound.

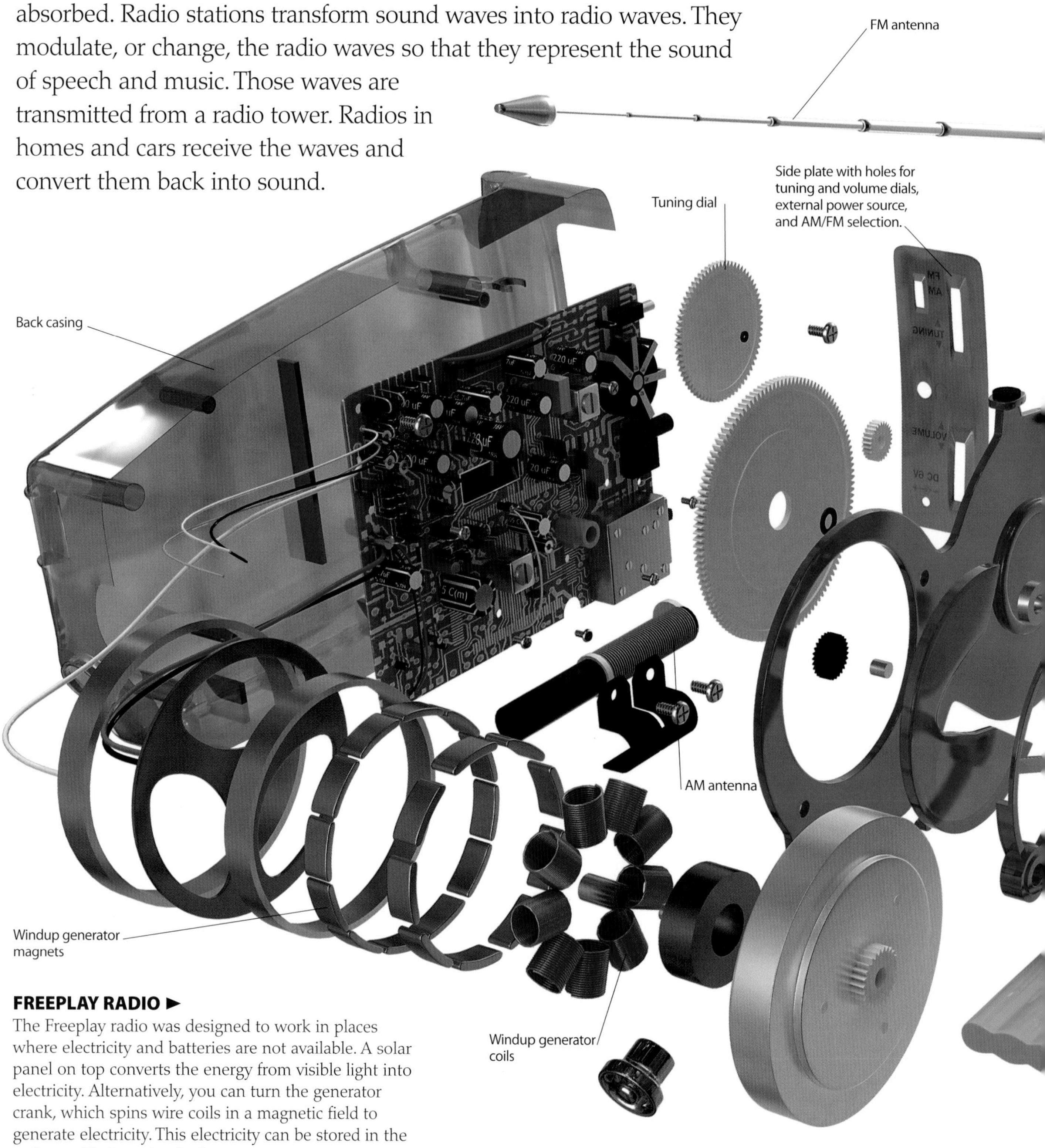

FM antenna

Side plate with holes for tuning and volume dials, external power source, and AM/FM selection.

Tuning dial

Back casing

AM antenna

Windup generator magnets

Windup generator coils

FREEPLAY RADIO ▶

The Freeplay radio was designed to work in places where electricity and batteries are not available. A solar panel on top converts the energy from visible light into electricity. Alternatively, you can turn the generator crank, which spins wire coils in a magnetic field to generate electricity. This electricity can be stored in the rechargeable batteries for later use.

TUNING IN ►

Radio announcers, news broadcasters, and disc jockeys typically have all or some parts of the broadcast programmed onto computers. The radio station broadcasts its radio waves at a particular frequency, or number of waves per second. When you tune your radio to that frequency, you hear that station's music. For example, if you listen to WROK 93.1 FM, the signal from the station is transmitted at 93.1 MHz (megahertz or millions of cycles per second). When you select 93.1 on your radio, the radio is receiving only radio waves at a frequency of 93.1 MHz, and you can hear your favorite station. AM stations broadcast at lower frequencies that are measured in KHz (kilohertz, or thousands of cycles per second).

did you know?
....
FM RADIO IS BROADCAST ONLY AT FREQUENCIES OF 88 MHZ TO 108 MHZ. AM RADIO IS BROADCAST ONLY AT 530 KHZ TO 1700 KHZ.

A microphone converts the sound waves of the broadcaster's voice into electrical energy.

The broadcaster uses the console to control the volume of her voice, the music, the advertisements, or any other sound that is being broadcast.

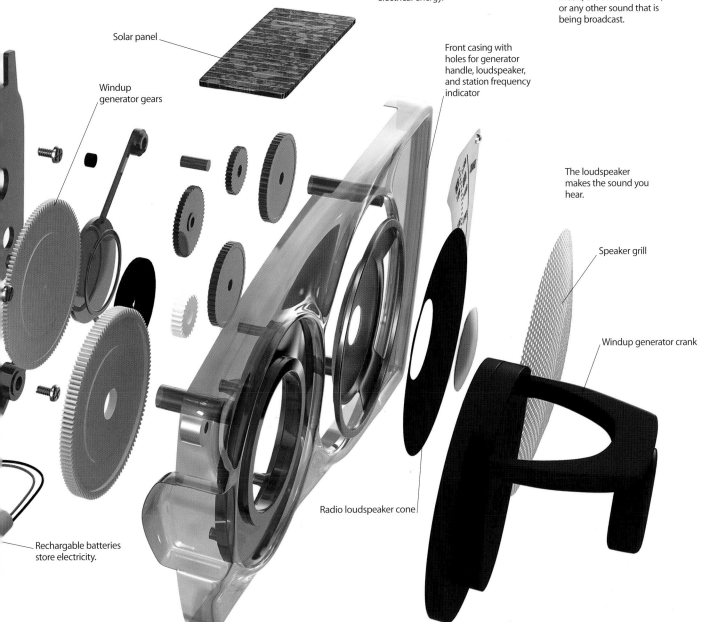

Solar panel

Windup generator gears

Front casing with holes for generator handle, loudspeaker, and station frequency indicator

The loudspeaker makes the sound you hear.

Speaker grill

Windup generator crank

Radio loudspeaker cone

Rechargable batteries store electricity.

GUITAR

A guitar's strings vibrate when you pluck, pick, or strum them. When you hear the sound, you know it is coming from a guitar—not a violin or a piano. An instrument's vibrations are waves of particular frequencies, the number of waves that occur in a certain period of time. Each note an instrument plays has a certain tone, called its *fundamental tone*. When you hear that tone from a guitar, which comes from a wave moving up and down a string, you are also hearing other tones, called *overtones*, that are characteristic of a guitar. The frequencies of the overtones are related to the fundamental tone mathematically—they are twice, three times, or another multiple of the frequency of the fundamental tone. You hear the sound as one note, but it is harmonious and musical because of its overtones.

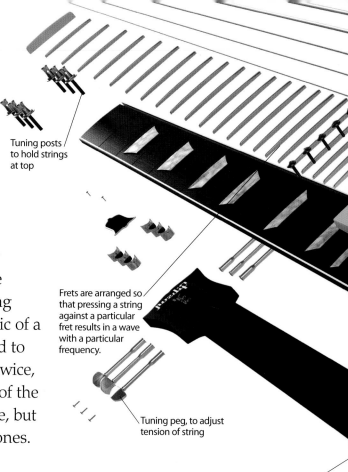

Tuning posts to hold strings at top

Frets are arranged so that pressing a string against a particular fret results in a wave with a particular frequency.

Tuning peg, to adjust tension of string

Neck

◄ UNPLUGGED
The strings of an acoustic guitar—one that is not electric—make the bridge vibrate. The bridge is the piece to which the strings are attached at the bottom. The bridge is attached to the flat front of the guitar—the piece with a hole in it—called the *soundboard*. The vibration of the strings transfers to the soundboard and then to the air inside the guitar, amplifying the sound.

did you know? A GUITAR CAN HAVE AS FEW AS 4 OR AS MANY AS 18 STRINGS.

Bridge

Soundboard

Pick guard

MUSICAL NOTATION ▲
Music is written on a framework of five lines, called a *staff* or *pentagram*. Notes are marked on the lines and spaces corresponding to the particular sound to be played. Notes are labeled A through G, and then back to A. The space from one A to the next A, or B to B, is called an *octave*. Written music is like a written language that uses notes instead of words.

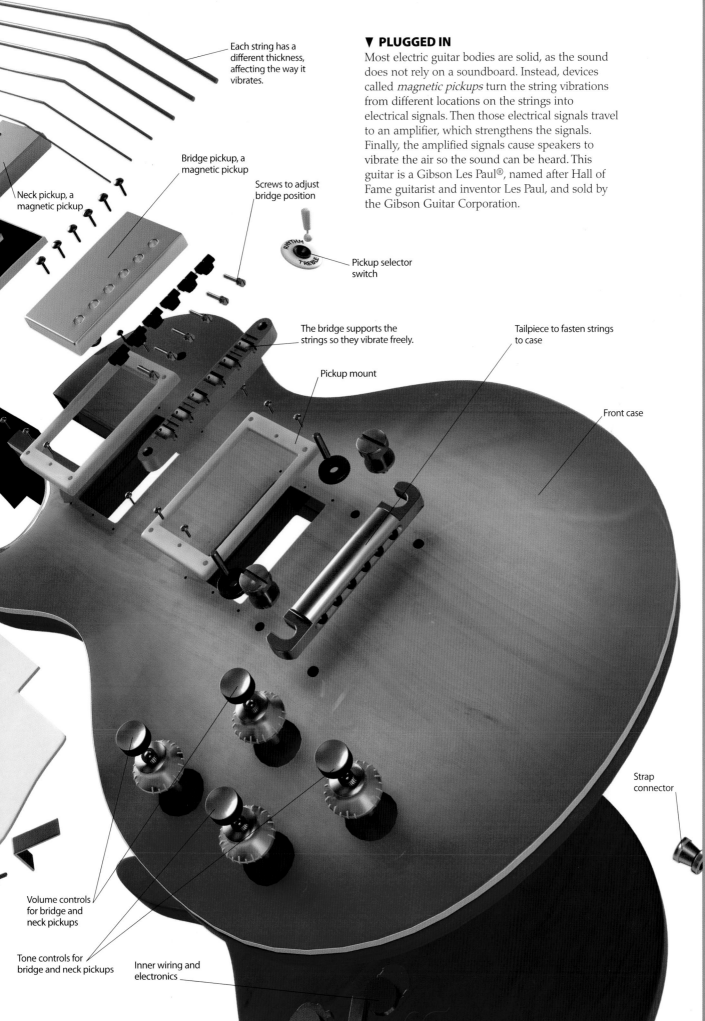

Each string has a different thickness, affecting the way it vibrates.

Bridge pickup, a magnetic pickup

Neck pickup, a magnetic pickup

Screws to adjust bridge position

Pickup selector switch

The bridge supports the strings so they vibrate freely.

Pickup mount

Tailpiece to fasten strings to case

Front case

Strap connector

Volume controls for bridge and neck pickups

Tone controls for bridge and neck pickups

Inner wiring and electronics

▼ PLUGGED IN

Most electric guitar bodies are solid, as the sound does not rely on a soundboard. Instead, devices called *magnetic pickups* turn the string vibrations from different locations on the strings into electrical signals. Then those electrical signals travel to an amplifier, which strengthens the signals. Finally, the amplified signals cause speakers to vibrate the air so the sound can be heard. This guitar is a Gibson Les Paul®, named after Hall of Fame guitarist and inventor Les Paul, and sold by the Gibson Guitar Corporation.

SONIC BOOMS

A jet plane speeds across the sky overhead and a sound like thunder follows. Boom! Was that the roar of its engines? Actually, it was a sonic boom. A sonic boom occurs whenever an object, such as the plane, travels faster than the speed of sound. To understand why a sonic boom occurs, think about sound as the movement of air molecules. When the plane moves through the air, it pushes the air molecules, producing pressure waves. These waves are alternating high pressure and low pressure areas of air. As the plane flies through the air, these waves travel in all directions including to the ground. You hear each wave as normal airplane noise. When the plane travels faster than the speed of sound, the waves produced by the plane can travel only as fast as the speed of sound, which is slower than the plane. Because of this, the waves "run into" each other and bunch up, forming very large waves, called *shock waves*. A shock wave is actually a sudden increase in pressure. When the shock waves reach your eardrum, you hear a sonic boom. The sudden increase in pressure has the same effect as the sudden expansion of air produced by an explosion. People nearby hear the sonic boom of the explosion when the shock wave passes them.

Vapor cone or condensation cloud

◄ VAPOR CONE

When a plane or rocket travels near the speed of sound, a cone-shaped cloud, called a *vapor cone*, may form around the object. This cloud forms due to a sudden acceleration of the airplane at high speeds, causing an abrupt drop in pressure of the air passing over the plane. When the air pressure decreases, the air temperature also decreases and water vapor in the air condenses into tiny droplets of water. Because this can happen to an object traveling just below the speed of sound, a sonic boom may not be heard.

FASTER THAN THE SPEED OF SOUND ▼

On October 14, 1947, Chuck Yeager, a test pilot with the United States Air Force, reached a speed of 700 miles per hour (about 1,126 km/h) or 1.06 times the speed of sound (also known as Mach 1.06). Yeager was flying a rocket-powered X-1 research plane. This plane, 31 feet (about 9.4 m) long with a wingspan of 28 feet (about 8.5 m), was modeled after the shape of a .50-inch (1.27-cm) bullet. With this feat, Yeager became the first person to travel faster than the speed of sound. To calculate an object's Mach number, you compare its speed with the speed of sound at the same temperature. For example, the speed of sound in air at Yeager's flying altitude of 40,000 feet (12,192 m) is about 660 miles per hour (1,062 km/h). When you divide Yeager's speed by the speed of sound—700/660 (1,126/1,062)—you get Mach 1.06.

Movement of the air around the cockpit of this F/A-18 Hornet has also caused a vapor cloud to form here, as well as behind the wings of the aircraft.

The pilot of this aircraft has a much greater visibility from the cockpit than Yeager had in the bullet-shaped X-1.

Concorde is 204 feet (about 62 m) long, increasing by 6 to 10 inches (about 15 to 25 cm) during flight. At supersonic speeds, friction with the air heats the exterior of the plane and the metal expands, so the plane lengthens at joints in the metal.

The wingspan of Concorde is 83 feet 8 inches (25.5 m).

A display showed Concorde passengers their speed and altitude while in flight. Mach 2 means that the plane is moving at twice the speed of sound. At an altitude of 54,000 feet (16,459 m), the speed of sound is about 660 miles per hour (1,062 km/h). That means the Concorde is flying at 1,320 miles per hour (2,124 km/h).

COMMERCIAL SUPERSONIC FLIGHT ▲

Flight that is faster than the speed of sound, called *supersonic flight*, is not just for the military. Fourteen Concordes®, supersonic commercial aircraft, carried more than 2.5 million people, starting in 1976 and ending in 2003. Concorde could fly as fast as 1,350 miles per hour (almost 2,173 km/h), traveling from London to New York in only 3.5 hours—half the time needed by standard aircraft.

FORMULA 1 CAR

At 200 miles per hour (about 322 km/h), a vehicle with wings could fly. Formula 1 cars reach that speed and do have wings attached to the front and back of the car. Their wings, however, are designed to create the opposite of lift, called *downforce*. This downward force allows cars to cling tightly to the track as they maneuver the tight curves of a Formula 1 race. However, this force also slows the car down. The other force slowing the car down is the force of the air pushing against the car as it moves forward. Every part of the car is designed to minimize these forces, by allowing air to move smoothly over, under, and around the car. The goal in designing these cars is to achieve the highest speed without flying off the track. This balance can get pretty tricky when 0.01 second can be the difference between winning and losing the race.

know?

did you

FORMULA 1 CARS PRODUCE SO MUCH DOWNFORCE THAT, AT 100 MILES PER HOUR (ABOUT 161 KM/H), THEY COULD THEORETICALLY BE DRIVEN UPSIDE DOWN ON A CEILING.

WINDING ALONG THE TRACK ▼

In the first automobile race, held in France in 1894, the average speed of the winning car was about 11 miles per hour (almost 18 km/h). Now around 20 Formula 1 races are held each year in countries around the world. The average speeds vary with the particular track, but on the straight portions, a Formula 1 car moves at least 180 miles per hour (almost 290 km/h)—much higher in some races.

GOING WITH THE FLOW

Wings, sometimes called *spoilers,* on the front and back of a Formula 1 car, are one of the most important design features. Their purpose is to direct air that is blasting against the car as it gains speed. Airplane wings are designed to direct the air that hits the wing downward, lifting the plane up. The rear wing on a Formula 1 car directs the air upward, so the back of the car is pressed down.

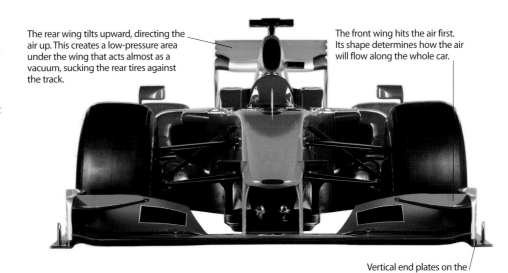

The rear wing tilts upward, directing the air up. This creates a low-pressure area under the wing that acts almost as a vacuum, sucking the rear tires against the track.

The front wing hits the air first. Its shape determines how the air will flow along the whole car.

Vertical end plates on the front wing reduce drag over the front tires.

The shape of the car body is rounded and decreases in width so that air flows over it with the least resistance.

The tires of a Formula 1 car have to be very light but still be able to handle a downforce of about 9,960 Newtons on each tire. Because of friction, the tires get as hot as boiling water.

FUEL CELL CARS

Everyday, millions of people add carbon dioxide, nitrous oxide, and other greenhouse gases—gases that trap heat in the atmosphere—to the air, simply by driving to the grocery store. In fact, the largest source of carbon dioxide, the most common greenhouse gas, is transportation. Earth needs some greenhouse gases. Without them, the temperature of the planet would be much colder. But human activities, such as driving, have led to an increase in greenhouse gases, ultimately leading to global warming. The temperature of Earth's surface has increased about 1°F (about 0.56°C) in the last 100 years. If we could eliminate greenhouse gases from vehicle emissions and replace them with water vapor, the warming effect on Earth could be greatly reduced. Engineers and scientists have developed fuel cell cars that do just that. The fuel is hydrogen and oxygen, and the product is electricity, heat, and water!

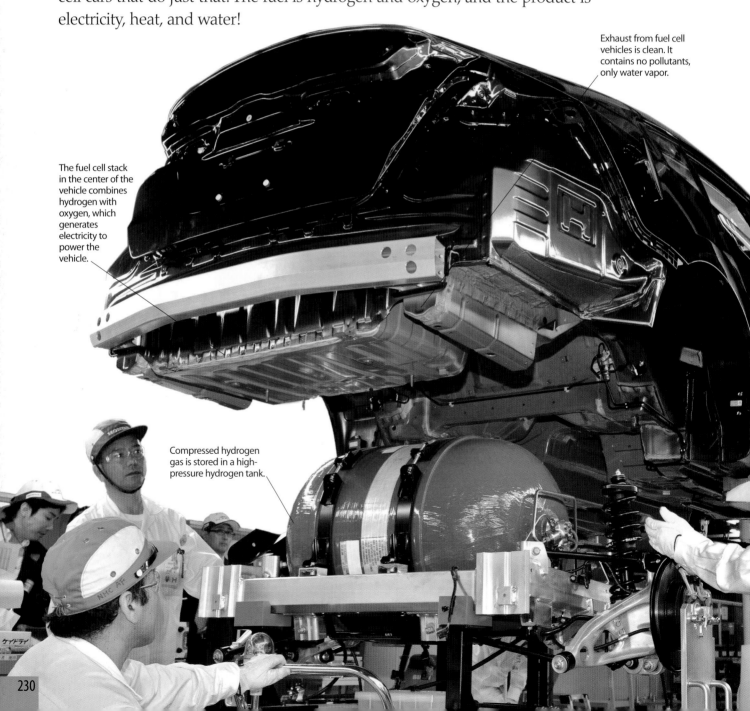

Exhaust from fuel cell vehicles is clean. It contains no pollutants, only water vapor.

The fuel cell stack in the center of the vehicle combines hydrogen with oxygen, which generates electricity to power the vehicle.

Compressed hydrogen gas is stored in a high-pressure hydrogen tank.

WHERE'S THE CLOSEST GAS STATION? ►

Some obstacles to getting fuel cell cars on the road is the lack of pipelines, trucks, and storage facilities to deliver the hydrogen to the fueling stations. One company that makes hydrogen fuel cell cars, is developing a home energy station that can produce hydrogen from natural gas. It would provide heat and electricity for the home and fuel for fuel cell vehicles. Using a home energy station would make carbon dioxide emissions 30 percent lower overall than those of the average household.

did you know?..................

FUEL CELLS POWER THE ELECTRICAL SYSTEMS OF NASA'S SPACE SHUTTLES. THE CREW DRINKS THE WATER PRODUCED BY THE CELLS.

◄ HYDROGEN FUEL CELL CAR

Hydrogen fuel cell cars are now available in places where hydrogen refueling stations exist. In Southern California, drivers can lease a car like the one shown here. The minivan shown above is being driven in Iceland, the location for some of the testing of hydrogen fuel cell vehicles.

HYDROGEN NOZZLE AND METER ►

Filling up your car with hydrogen is slightly more complicated than filling it up with gasoline. Most fuel cell vehicles use gaseous hydrogen. The pump at the gas station has to read the pressure in the gas tank and determine how much fuel is needed. The nozzle is also different. It locks into place so that hydrogen cannot escape.

Nozzle

FUEL CELL HYBRID ▼

One company has created this fuel cell hybrid vehicle (FCHV), which is based on a popular SUV model. It is able to run either on electricity or fuel cells. The vehicle is being tested, but is not in production yet.

The electric drive motor of hydrogen fuel cell cars is exceedingly quiet, reducing noise pollution and making for a quiet, smooth ride.

CREATING ELEMENTS

Scientists describe the origin of the universe as a sudden expansion of matter and energy—the big bang. Particles formed, and then joined to create some of the elements. Three minutes after the big bang, most of the hydrogen that exists today had formed. It took a while longer to make helium and traces of lithium—maybe 35 minutes or so. The rest of the elements were formed in the stars. Elements are made in the stars through nuclear fusion, which is the formation of heavier elements from lighter ones. As a star burns its fuel, gravity pulls its material inward and it gets hotter—hundreds of millions of degrees hotter. Then, atoms collide and fuse to make the heavier elements. It takes the intense heat of supernovas to make elements heavier than iron. A supernova is an explosion of a huge star. The pieces flung out in this explosion come together to create new stars and planets. That's how elements that formed in the stars came to exist on Earth.

THE PERIODIC TABLE ▶

The periodic table is a tool that people use to organize the elements. Each element has a unique identity, determined by the number of protons in its nucleus—called its *atomic number*. Atomic numbers increase from left to right in each row. The elements in the same column have similar chemical and physical properties. The table shows each element's symbol, which is a one- or two-letter abbreviation.

Hydrogen is the lightest and most abundant element in the universe. It is the only element on the left side of the table that is not a metal.

did you know? MORE THAN 40 ELEMENTS ARE FOUND IN THE HUMAN BODY, BUT CARBON, OXYGEN, HYDROGEN, AND NITROGEN MAKE UP 96 PERCENT OF OUR CELLS.

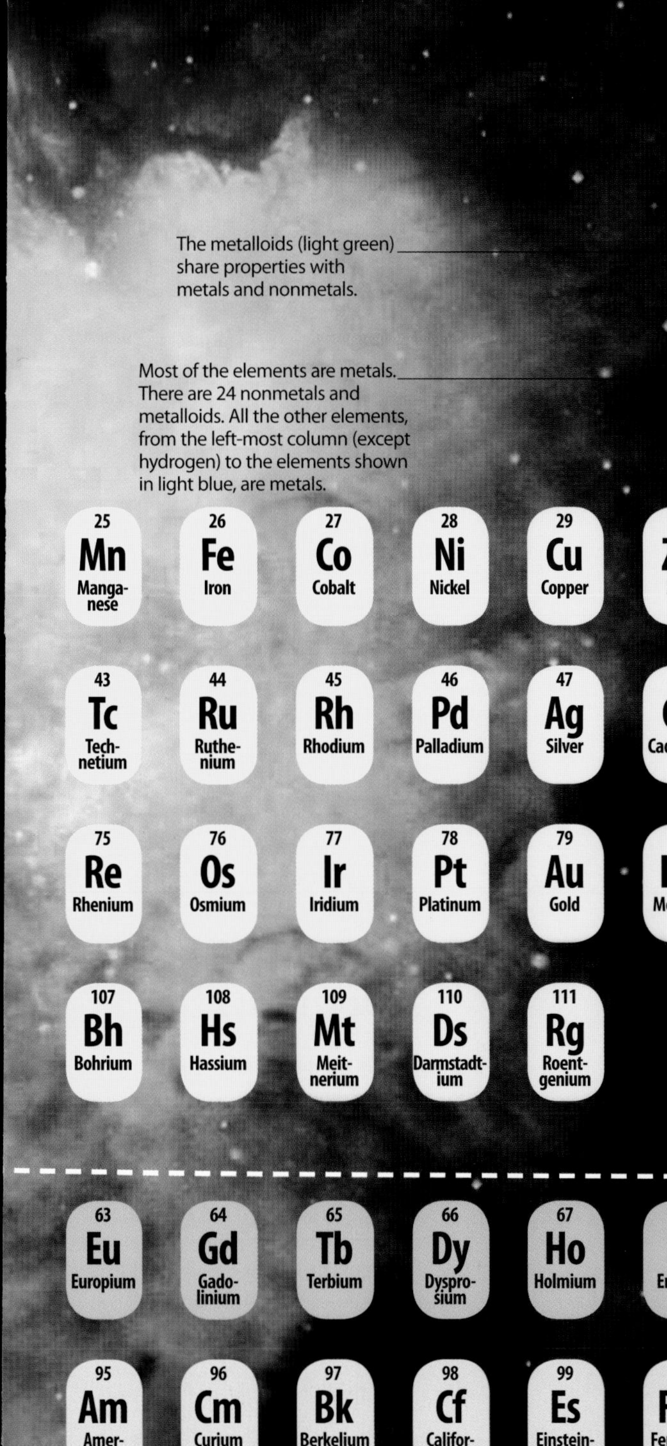

◄ **BIRTHPLACE OF STARS**

A supernova is essentially the death of a star. It blows the outer layers of a star far into space. Its matter mixes with interstellar gases, mostly hydrogen, forming a huge cloud of dust and gas called a *nebula*, like the one shown here. Within the nebula, gravity pulls bits and pieces together, forming new stars. Part of the material forms planets, such as Earth, whose core is mostly iron.

Along with hydrogen, elements shown in green and blue, to the right of the metalloids, are nonmetals. Their properties are very different from those of the metals.

| 2 | He | Helium |

The metalloids (light green) share properties with metals and nonmetals.

Most of the elements are metals. There are 24 nonmetals and metalloids. All the other elements, from the left-most column (except hydrogen) to the elements shown in light blue, are metals.

| 5 B Boron | 6 C Carbon | 7 N Nitrogen | 8 O Oxygen | 9 F Fluorine | 10 Ne Neon |

| 13 Al Aluminum | 14 Si Silicon | 15 P Phosphorus | 16 S Sulfur | 17 Cl Chlorine | 18 Ar Argon |

| 25 Mn Manganese | 26 Fe Iron | 27 Co Cobalt | 28 Ni Nickel | 29 Cu Copper | 30 Zn Zinc | 31 Ga Gallium | 32 Ge Germanium | 33 As Arsenic | 34 Se Selenium | 35 Br Bromine | 36 Kr Krypton |

| 43 Tc Technetium | 44 Ru Ruthenium | 45 Rh Rhodium | 46 Pd Palladium | 47 Ag Silver | 48 Cd Cadmium | 49 In Indium | 50 Sn Tin | 51 Sb Antimony | 52 Te Tellurium | 53 I Iodine | 54 Xe Xenon |

| 75 Re Rhenium | 76 Os Osmium | 77 Ir Iridium | 78 Pt Platinum | 79 Au Gold | 80 Hg Mercury | 81 Tl Thallium | 82 Pb Lead | 83 Bi Bismuth | 84 Po Polonium | 85 At Astatine | 86 Rn Radon |

| 107 Bh Bohrium | 108 Hs Hassium | 109 Mt Meitnerium | 110 Ds Darmstadtium | 111 Rg Roentgenium |

The metals beneath this line are two groups of chemically similar elements. They are almost always set apart so that the table will fit across a page.

| 63 Eu Europium | 64 Gd Gadolinium | 65 Tb Terbium | 66 Dy Dysprosium | 67 Ho Holmium | 68 Er Erbium | 69 Tm Thulium | 70 Yb Ytterbium |

| 95 Am Americium | 96 Cm Curium | 97 Bk Berkelium | 98 Cf Californium | 99 Es Einsteinium | 100 Fm Fermium | 101 Md Mendelevium | 102 No Nobelium |

KEY TO ELEMENT COLORS

- Alkali metals
- Alkaline earth metals
- Transition metals
- Lanthanides
- Actinides
- Metals in mixed groups
- Metalloids
- Nonmetals
- Noble gases

QUARKS AND LEPTONS

What are you made of? You might think of bones, blood, skin, hair, or cells. But if you look closer, what are cells made of? For thousands of years, people have searched for the basic building blocks of matter. For the past 200 years, this search has focused on atoms and the particles that make up atoms: protons, neutrons, and electrons. Now, physicists think they have identified the fundamental particles—the smallest pieces of matter that explain what makes up the universe and what holds it together. Physicists have identified 6 types of particles called *quarks*, and 6 called *leptons*. An electron is one type of lepton. Protons and neutrons, however, are made up of quarks. Physicists have developed some interesting ways of describing the categories of quarks. The 6 quarks are one of 6 "flavors": up, down, top, bottom, charm, or strange. And each is further divided into one of 3 "colors": red, blue, or green. The discovery of quarks and leptons has posed many new questions and hypotheses for scientists to answer.

Stomach

▲ MUSCLE CELLS

Every living organism, including you, is made of cells. A human body is made of thousands of different types of cells working together. Within the stomach, for example, are muscle cells that work together, moving food through the digestive system. When these cells contract, they mash food and mix in enzymes that help break down larger molecules into smaller ones.

◄ MATTER EVERYWHERE

The girl and her tennis racket are both examples of matter. All living organisms and nonliving matter are made up of the fundamental particles called quarks and leptons.

▼ HOLDING THE CELL TOGETHER

A portion of the structure known as the cell membrane is shown below. The cell membrane separates the inside of the cell from its surroundings. The cell membrane holds the cell together and controls the movement of molecules into and out of the cell. A cell membrane is made up of molecules that include proteins and phospholipids. Phospholipids form a double layer, and they consist of groups of atoms that are arranged in shapes that look something like a head and two tails.

did you know? MOST COMMON MATTER CONSISTS OF UP QUARKS AND DOWN QUARKS, NOT THE OTHER 4 KINDS OF QUARKS.

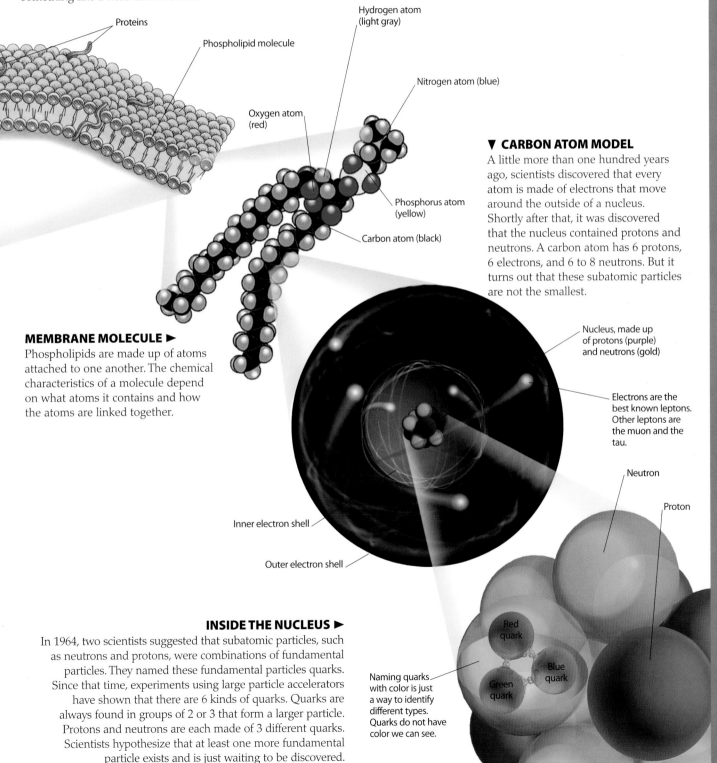

Proteins

Phospholipid molecule

Hydrogen atom (light gray)

Nitrogen atom (blue)

Oxygen atom (red)

Phosphorus atom (yellow)

Carbon atom (black)

▼ CARBON ATOM MODEL

A little more than one hundred years ago, scientists discovered that every atom is made of electrons that move around the outside of a nucleus. Shortly after that, it was discovered that the nucleus contained protons and neutrons. A carbon atom has 6 protons, 6 electrons, and 6 to 8 neutrons. But it turns out that these subatomic particles are not the smallest.

MEMBRANE MOLECULE ▶

Phospholipids are made up of atoms attached to one another. The chemical characteristics of a molecule depend on what atoms it contains and how the atoms are linked together.

Nucleus, made up of protons (purple) and neutrons (gold)

Electrons are the best known leptons. Other leptons are the muon and the tau.

Neutron

Proton

Inner electron shell

Outer electron shell

Red quark

Green quark

Blue quark

INSIDE THE NUCLEUS ▶

In 1964, two scientists suggested that subatomic particles, such as neutrons and protons, were combinations of fundamental particles. They named these fundamental particles quarks. Since that time, experiments using large particle accelerators have shown that there are 6 kinds of quarks. Quarks are always found in groups of 2 or 3 that form a larger particle. Protons and neutrons are each made of 3 different quarks. Scientists hypothesize that at least one more fundamental particle exists and is just waiting to be discovered.

Naming quarks with color is just a way to identify different types. Quarks do not have color we can see.

CRYSTALS

Crystals form amazing shapes because their atoms and molecules are bonded together in an orderly, regular, repeated pattern. The pattern gives them their straight edges and smooth faces. But crystals aren't just flashy gems. Everyday things, such as sugar and salt, are also crystals. Scientists classify crystals in many different ways. One is based on the way their molecules or atoms are bonded. Some crystals are molecules that are weakly bonded together, such as ice, or rock candy made from sugar. Salt, an ionic crystal, has stronger bonds, because its atoms are bonded by the attraction of oppositely charged particles called *ions*. Salt dissolves in water, but if the water evaporates, crystals of salt form again. In metallic crystals, atoms are packed tightly to form a highly dense structure. Diamonds are called *covalent* crystals, because their atoms are bonded by sharing an electron. These bonds are extremely strong.

Axinite crystals are flattened and wedge-shaped.

did you know?

THE WORLD'S LARGEST CRYSTALS ARE ALMOST 40 FEET (ABOUT 12 M) LONG AND WEIGH ALMOST 60 TONS. THEY ARE GYPSUM CRYSTALS, LOCATED DEEP WITHIN THE CAVE OF CRYSTALS IN MEXICO.

PLEASE PASS THE SALT ▼

Southern France is famous for its seriously salty scenery. Located on the Mediterranean coast, the Camargue region is filled with inland salt lagoons that evaporate in the summer sun, leaving large piles of sea salt behind. The salt piles form huge mounds, some of which can grow up to 26 feet (about 8 m) tall. These salt mounds support a thriving salt industry in Camargue. Now, if only they had a pretzel factory nearby!

CRYSTAL SYSTEMS ▶

Crystals can be categorized by the arrangement of their atoms, called *lattices*. Imagine that a baseball represents an atom. If you arrange baseballs into a cube, you form what's called a *unit cell*. If you stack this cube onto another cube, you form a lattice. The shape of the unit cell is important. For example, both graphite and diamond consist only of carbon atoms. But because graphite and diamond have different shaped unit cells, they have different properties. Graphite is black and soft enough to lubricate plastic, metal, or wood, while diamond is clear and hard enough to cut many materials, including glass. The drawings on the right show various shapes of unit cells.

Cubic (or Isometric): halite (rock salt), diamond, pyrope (a type of garnet), spinel

Orthorhombic: aragonite, olivine, topaz, sulfur, tanzanite

Hexagonal/Trigonal (two similar systems): graphite, forms of beryl (aquamarine and emerald); forms of quartz (amethyst, ruby, sapphire)

Tetragonal: zircon, cassiterite (tin oxide), rutile (titanium oxide)

Triclinic: axinite, turquoise, rhodonite, wollastonite

Monoclinic: gypsum, malachite, talc, muscovite, azurite

Aquamarine crystals are used as gemstones.

Talc crystals break off in flakes that feel soapy.

Topaz crystals are shaped like prisms or double pyramids. They come in a variety of colors.

FLUORESCENT MINERALS

If you look closely at many different rocks, you will notice that some contain shiny crystals, or minerals. Minerals can form as molten rock cools, causing atoms to bond together. They can also form when water that contains dissolved elements evaporates, leaving the elements behind. Often the atoms of these elements are close enough to bond together to form crystals. A pure mineral is made up of a single compound and is usually colorless. Add a tiny amount of another element, or an impurity, and you can get some amazing effects. The brilliant green of an emerald and red of a ruby come from crystal impurities. Some minerals even seem to glow. Fluorescent minerals shine brightly when they are exposed to ultraviolet light, sometimes called *black light*. Many natural history museums have collections of these minerals. Under normal lighting, you see interesting shapes, often gray or brown in color. When the black light comes on, though, the room shines with bright yellows, reds, greens, and blues as the minerals give off these colors.

You often find several minerals together. Under the UV light, it is obvious that the quartz crystals surrounding the calcite are made of a different material.

238

▼ GLOWING IN THE DARK

Under normal lighting, this large calcite crystal appears as a white hexagonal column. This crystal contains an impurity—manganese ions. When the crystal is exposed to ultraviolet (UV) light, the electrons in the manganese absorb energy. Then they give off this energy in the form of reddish-orange light.

Fluorescence is a type of luminescence, light that is produced without heat. The energy comes from ultraviolet light.

GLOWING IN THE LIGHT ▲

The word *fluorescence* comes from the mineral fluorite, the first fluorescent mineral to be discovered. This blue color, caused by europium atoms in the crystal, is sometimes bright enough that the crystal glows in sunlight.

Different impurities produce different colors. Uranium in quartz produces green light, while mercury can produce pink or bright blue light.

did you know? THE FLUORESCENT MINERAL CAPITAL OF THE WORLD IS THE BOROUGH OF FRANKLIN IN NORTHERN NEW JERSEY.

MELTING POINT

Knowing the melting point of a solid—the temperature at which it becomes a liquid—helps you make decisions all the time. You put bread in a toaster, but you don't insert a bar of chocolate. You bake cookies on a sheet made of metal, not plastic. Generally, you expect solids to stay solid. You take for granted that your lunch will not melt in your backpack and your bicycle will not melt in the sun. Scientists use the melting point of a substance as a way to identify an unknown chemical or to determine which material to use for a particular task. A scientist can determine each ingredient in a tablet or pill by measuring the temperature at which each substance melts. To make clothing, utensils, electronic devices, and most everything we use each day, manufacturers choose the material based partly on its melting point.

Chocolate starts out as a thick liquid made of roasted, ground cocoa beans. It is molded into solid shapes as it cools.

MELTS IN YOUR MOUTH ▶

Chocolate is made up of about half fat, called *cocoa butter*, and half cocoa particles—similar to dry cocoa powder. When you melt chocolate, the cocoa butter melts. The change is physical, because the cocoa butter will harden again at room temperature. The melting point of chocolate is just below your body temperature of around 98.6°F (37°C). When the chocolate melts on your tongue, it absorbs thermal energy from your mouth, which is an endothermic change.

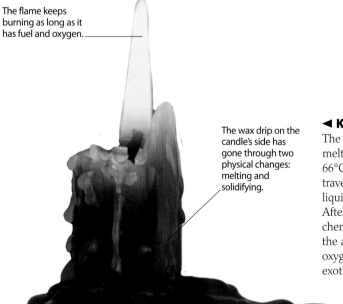

The flame keeps burning as long as it has fuel and oxygen.

The wax drip on the candle's side has gone through two physical changes: melting and solidifying.

did you know? ....
TUNGSTEN METAL HAS THE HIGHEST MELTING POINT OF THE METALLIC ELEMENTS: 6,152°F (3,410°C). IT IS USED TO MAKE THE THIN WIRE THAT GLOWS IN LIGHT BULBS.

◀ KEEPING A FLAME BURNING

The flame of a candle produces plenty of heat to cause wax to melt. Candle wax melts at a temperature of about 150°F (about 66°C) and the flame is around 2,552°F (1,400°C). The liquid wax travels up the candle's wick. The flame heats the molecules of liquid wax so that they vibrate fast enough to vaporize into a gas. After these two physical changes—melting and vaporizing—a chemical change takes place. The chemical bonds between the atoms of the gaseous wax break. The atoms bond with the oxygen in the air, and combustion, or burning, takes place. This exothermic reaction releases energy in the form of heat and light.

Most iron is found in iron ore, which consists of rocks and minerals surrounding the iron.

The material used to hold the molten iron has to have a melting point higher than that of iron.

The melting point of iron is 2,800°F (about 1,538°C).

MOLTEN IRON ►

Iron and other metals can be melted, poured into molds, and cooled. These are physical changes. Other molten metals can be added to molten iron to produce solid solutions called *alloys*. The atoms of the two metals remain unchanged, so making an alloy is a physical change. Combining small amounts of carbon and other elements with iron makes various types of steel, which is an alloy. This combining can add certain desirable qualities to the alloy, such as strength or flexibility.

Molten iron is poured into a mold to solidify. The solid piece of iron is called an *ingot*.

GLASS

People have been making glass for thousands of years. The earliest known glass objects were beads, made by Egyptians around 3,500 B.C. In about 27 B.C., Syrians learned they could insert a long metal tube into molten glass and blow into it to create hollow glass containers, such as vases and bottles. Glass is an unusual material. When it is a hot, molten liquid, it can be formed into shapes. Then it hardens into a transparent solid but still retains some properties of a liquid. When most substances are solid, the molecules are like tightly packed bricks in a wall. The molecules of liquids are farther apart, allowing light to pass through. This is why most liquids are clear or semi-clear and most solids are opaque. Glass has properties of both liquids and solids. The molecules in a glass window, for example, do not move, but are far enough apart to allow light to pass through.

MAKING LIQUID GLASS ▼

People make glass by melting pure sand with other minerals in a furnace heated to 3,092°F (1,700°C). They add soda ash to lower the melting point—the temperature at which solids become liquid. Limestone is added to increase the strength and stability of the glass. Adding broken pieces of other glass speeds up the melting process. Ingredients such as copper, gold, and other chemical elements give glass a variety of colors.

72 percent sand

The remaining 5 percent can include chemicals that affect the color of the glass. Iron oxide is used in green or brown glass. Crystal glass and television glass contain barium carbonate.

15 percent soda ash (sodium carbonate)

8 percent limestone (calcium carbonate)

◀ GLASS BLOWING

A glassworker winds the glass, in its liquid state, onto a long, hollow iron rod and blows a bubble of air into the glass to give it a pear shape. Then, the worker rolls the iron rod shaping the glass. Last, the glass blower reheats the glass and blows more air into it to give the object its final shape.

It takes a worker years of practice to know when glass is the right consistency to blow and shape.

A glass blower shapes glass into a vase.

Blue glass

Patterned glass

Green glass

THE COLORS OF GLASS ▲

Cobalt turns glass dark blue, while gold can make glass ruby red. A small amount of chromium turns glass emerald green. Glass made from beach sand is usually light green or blue, since beach sand often has impurities, such as iron, in it. Lead makes glass sparkle and reflect.

243

AEROGELS

This block of material is nicknamed "frozen smoke," but it is no gas. It's aerogel—a solid form of silica, the same compound that makes up glass and sand. One of aerogel's amazing properties is that it traps heat, so well that a block of aerogel can protect a human hand from the flame of a blowtorch. Aerogels are composed mostly of air. They start out as wet gels that are much like the gelatin that you eat for dessert. The wet gel is then dried under high temperatures and pressures. These conditions vaporize the liquid, leaving what is called a *matrix*. This is a jungle-gym-like network of silica molecules surrounding microscopic pockets of air. Aerogels are more than 90% air, making them excellent thermal insulators. A thermal insulator is a material that is a poor conductor of thermal energy. The trapped air inside the aerogel slows the transfer of heat so the heat stays in the aerogel instead of passing through it.

◄ WHITE OR CLEAR?

The smoky, white-blue appearance of aerogels does not come from color within the materials that make them up. Instead, the color comes from the scattering of light by the pockets of air trapped inside the matrix. Polar bear fur looks white for the same reason. There are no white pigments, or color molecules, in the fur. Each hair is a transparent, hollow tube filled with air. This trapped air is an effective thermal insulator, keeping these animals warm in their frigid environment.

▼ EFFECTIVE INSULATOR

If you want to keep your ice cream chilly and your house toasty warm, you have to slow the transfer of heat, or thermal energy, with a thermal insulator. Many types of thermal insulators—such as the foam panels that line a freezer and the fiberglass mats that fill the walls of many homes—rely on trapped air to slow the transfer of heat. Aerogels are used to insulate some extreme-weather jackets, blankets, sheets, windows, and skylights.

did you know? ASTRONOMERS USE AEROGEL AS "SPONGES" TO CAPTURE THE DUST FROM PASSING COMETS.

The flame of the blowtorch is hot enough to heat and melt the surface of the aerogel, but heat does not conduct far enough into the block to reach the hand.

The flame of a blowtorch can be higher than 2,300°F (1,300°C).

Because aerogels are made mostly of air, they also make great sponges. Researchers are looking into ways of using aerogels as water filters and even as sponges to soak up oil spills.

Aerogels hold the world record for the least dense solid—just slightly denser than air—yet they can support objects many times their own mass.

STEEL

Iron, aluminum, magnesium, chromium. These are just a few of the elements in the periodic table known as metals. Most metals are in the form of a metal crystal. These crystals consist of closely packed, positively charged metal atoms, called *ions*, with negatively charged electrons drifting among them. This structure makes metal a good conductor of electricity and heat, and makes it possible to bend and shape metals without breaking them. However, most of the metals we use regularly are not single elements, but combinations of elements, called *alloys*. Steel is one of the most widely used alloys. It is mostly iron, a strong and plentiful metal. However, when first extracted from iron ore, crude iron (also called *pig iron*) is brittle. Adding small amounts of carbon increases the metal's strength and hardness. These steel alloys are called *carbon steel*. Other elements added to steel provide additional desirable characteristics. For example, adding chromium to steel provides resistance to rust and scratches.

EIFFEL TOWER ▼

The Eiffel Tower, a famous landmark in Paris, France, is made of puddle iron, not steel. Puddle iron is a type of wrought iron that was used in construction before steel. This low-carbon iron can be made by melting and combining iron ore with carbon. The resulting iron is strong and malleable, which means it can be rolled or hammered into sheets or bars.

WELDING ▶

Steel parts can be joined with nuts and bolts or with thick metal pins called *rivets*. However, welding is an effective and commonly-used way to join certain types of steel. Welding typically uses heat to melt the materials to be joined. An additional material may be used to join the parts. When cool, these parts are connected with a very strong bond.

◀ **COILED STEEL**

Steel, like all metals, is ductile, which means it can be pulled into long, thin, strong strands of wire. To make steel cable, these thin strands are twisted together to form a thicker strand. The thicker strands are combined to make a cable. Steel cables can support tremendous weight. For example, each of the four 15.75-inch-diameter (40 cm) steel suspension cables on the Brooklyn Bridge can support up to 24,621,780 pounds (11,168,252 kg).

The steel typically used in construction, called low carbon steel, contains between 0.05 and 0.30 percent carbon.

As an alloy of iron, steel corrodes less than pure iron would under similar circumstances.

The high heat at this point melts the steel, forming a bond that can be as strong as the steel itself.

FIREWORKS

If fireworks sound like gunshots to you, it might be because they rely on the same chemical reaction—the burning of black powder. Black powder is another name for gunpowder. It is a mixture of sulfur, charcoal (primarily carbon), and saltpeter (a chemical that contains potassium, nitrogen, and oxygen). When lit, these chemicals react explosively, leading to a big bang, a bright light, lots of heat, and sometimes, injuries. The explosiveness of black powder compared with that of regular charcoal has to do with the way charcoal burns under different conditions. When charcoal is burned alone, it reacts slowly with the small supply of oxygen in the air around it. However, when mixed with the other ingredients of black powder, charcoal can react much more quickly with the oxygen in the saltpeter. The result is a fast reaction that can be used to create the exciting (and loud) displays of a fireworks show.

◄ FIRECRACKERS

Firecrackers are simply small packages of black powder that make a big bang when lit. Some come as a string of firecrackers, having short fuses that are attached to a long fuse. This arrangement lets people light only the long fuse to set off several firecrackers in a row. Firecrackers are illegal in some states.

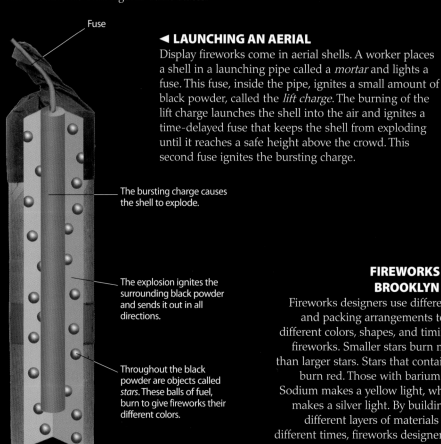

Fuse

◄ LAUNCHING AN AERIAL

Display fireworks come in aerial shells. A worker places a shell in a launching pipe called a *mortar* and lights a fuse. This fuse, inside the pipe, ignites a small amount of black powder, called the *lift charge*. The burning of the lift charge launches the shell into the air and ignites a time-delayed fuse that keeps the shell from exploding until it reaches a safe height above the crowd. This second fuse ignites the bursting charge.

The bursting charge causes the shell to explode.

The explosion ignites the surrounding black powder and sends it out in all directions.

Throughout the black powder are objects called *stars*. These balls of fuel, burn to give fireworks their different colors.

FIREWORKS OVER THE BROOKLYN BRIDGE ►

Fireworks designers use different materials and packing arrangements to design the different colors, shapes, and timing effects of fireworks. Smaller stars burn more quickly than larger stars. Stars that contain strontium burn red. Those with barium burn green. Sodium makes a yellow light, while titanium makes a silver light. By building stars with different layers of materials that burn at different times, fireworks designers can create dramatic color changes.

did you know?.........................
FIREWORKS BURN AT TEMPERATURES GREATER THAN 3,600°F (2,000°C). THIS TEMPERATURE IS NEARLY TWICE AS HOT AS A CHARCOAL FIRE.

LICHTENBERG FIGURES

During a thunderstorm, an electric charge builds up in the clouds and the ground. Suddenly, a flash of lightning pierces the sky in a jagged, branching line. Lightning lasts only a fraction of a second, but there is a way to "capture" the jagged traces of moving electrons in plastic. The Lichtenberg figures shown here were made by building a huge electric potential, called *voltage*, inside a plastic block. When the voltage gets high enough, the electrons move. In less than a millionth of a second, channels form inside the block as the electrons tear apart the chemical bonds of the plastic. The electrons that were trapped in the block rush out through the plastic, releasing the charge. Lichtenberg figures are named after the German physicist who discovered them in the 1700s. When he exposed insulating materials to high voltage, he saw branching images on the surface. Today, we use his discovery in printers and copy machines when charged surfaces pick up toner and put it on paper.

As more electrons are trapped in one place, they repel one another. The greater the electric field, the harder they push.

Conducting paths form when the electrons break apart chemical bonds. Lightning follows a similar path as gas molecules in the air are torn apart.

ELECTRICAL FRACTALS ►

A fractal is a geometric form in which a pattern is repeated at smaller and smaller scales. This Lichtenberg figure is a good example of a fractal. Look at the widest lines and how they branch. Each smaller branch breaks apart in the same way. Follow the track and you see another branch with similar proportions and angles. There are fractal patterns in nature. Upside down, this figure resembles the branches of a tree.

◄ LIGHTNING IN A CUBE

A particle accelerator pumped high-energy electrons into this plastic block. Because the plastic does not conduct electricity, the electrons were trapped in place, building energy as the electric field grew to millions of volts. Eventually, they had to move. Suddenly, electrons started tearing chemical bonds apart, creating charged channels. With a bang and a flash, electrons rushed away from one another and formed a Lichtenberg figure.

did you know?..............................

LICHTENBERG FIGURES SOMETIMES APPEAR ON THE SKIN OF PEOPLE WHO ARE STRUCK BY LIGHTNING, THOUGH THEY FADE WITHIN DAYS OR HOURS.

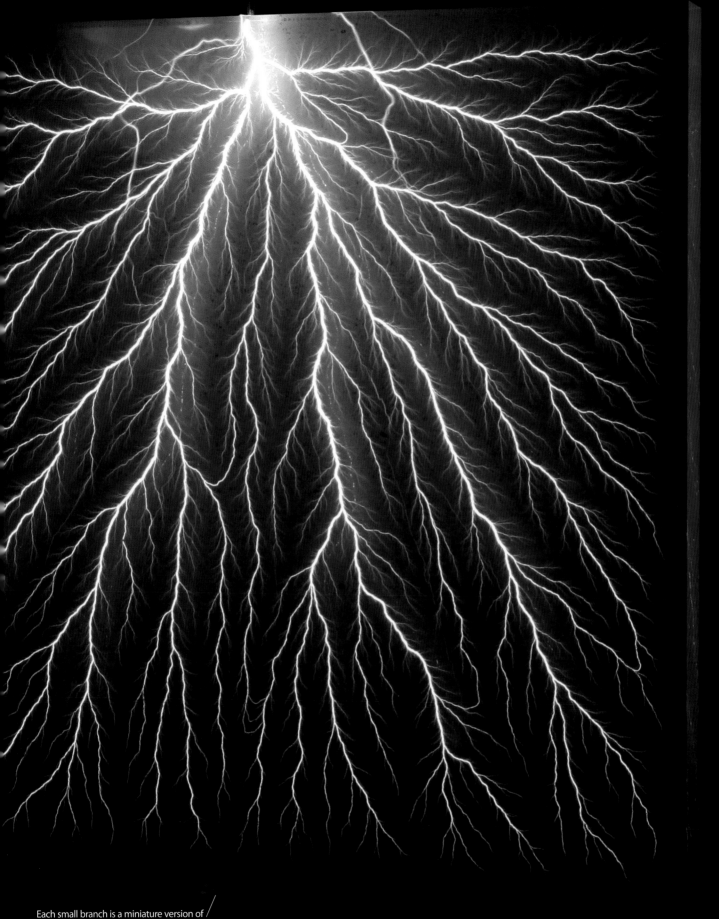

Each small branch is a miniature version of the larger branch. As electrons rush apart, they zigzag away from one another.

MICROSCOPES

Most people can see something as small as a human hair. But if you want to examine the tiny hairs on an ant, you need a microscope. A microscope is a tool used to see things that are too small to see with your eyes alone. The microscopes in most science classrooms are light microscopes. They focus light using curved pieces of glass, or lenses. Light microscopes are useful for viewing objects only as small as a cell. To see something smaller, like an atom, you need an electron microscope. An electron microscope uses a beam of electrons to magnify a sample. Transmission electron microscopes (TEMs) detect how electrons interact with the sample as they travel through it. They produce two-dimensional views that look like cross sections. Scanning electron microscopes (SEMs) detect electrons that bounce off the sample's surface, which has been covered with a thin layer of metal, usually gold. They produce three-dimensional images.

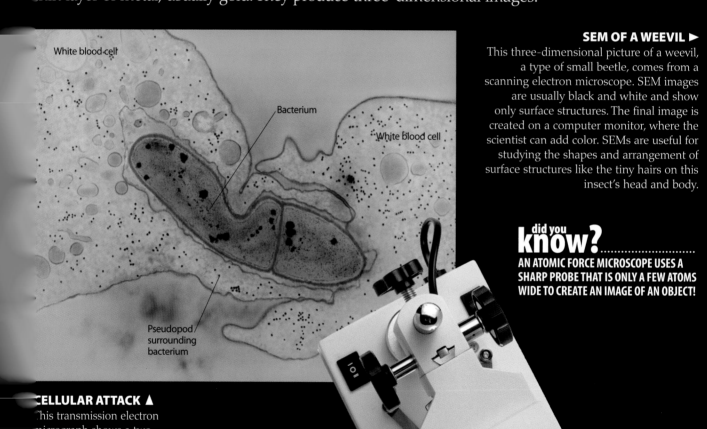

White blood cell

Bacterium

White blood cell

Pseudopod surrounding bacterium

SEM OF A WEEVIL ▶

This three-dimensional picture of a weevil, a type of small beetle, comes from a scanning electron microscope. SEM images are usually black and white and show only surface structures. The final image is created on a computer monitor, where the scientist can add color. SEMs are useful for studying the shapes and arrangement of surface structures like the tiny hairs on this insect's head and body.

did you know?..........................
AN ATOMIC FORCE MICROSCOPE USES A SHARP PROBE THAT IS ONLY A FEW ATOMS WIDE TO CREATE AN IMAGE OF AN OBJECT!

CELLULAR ATTACK ▲

This transmission electron micrograph shows a two-dimensional, cross-section view of two white blood cells (blue blobs coming together). They are capturing and digesting an invading bacterium (purple). This type of bacterium can cause some types of food poisoning. Notice the white blood cells' pseudopods. These armlike extensions of the cell membrane surround the unwanted bacterium and destroy it.

LIGHT MICROSCOPE ▶

Light microscopes use multiple lenses to magnify objects up to 1,000 times their original size. Scientists use them to study both living and dead specimens, such as slices of plant and animal tissues or drops of water that contain microorganisms, cells, or other small structures. Specimens are placed on clear glass slides. Light shining from below or above illuminates them.

ENVIRONMENTAL SCANNING ELECTRON MICROSCOPE (ESEM) ▶

Because conventional SEMs require that the sample be coated in gold and examined under a vacuum, samples cannot be alive. The environmental scanning electron microscope (ESEM), however, overcomes these problems. Only the narrow electron path of the sensor is under a vacuum. ESEMs can produce three-dimensional images of living things without killing them, coating them, or permanently changing them.

UNIVERSE

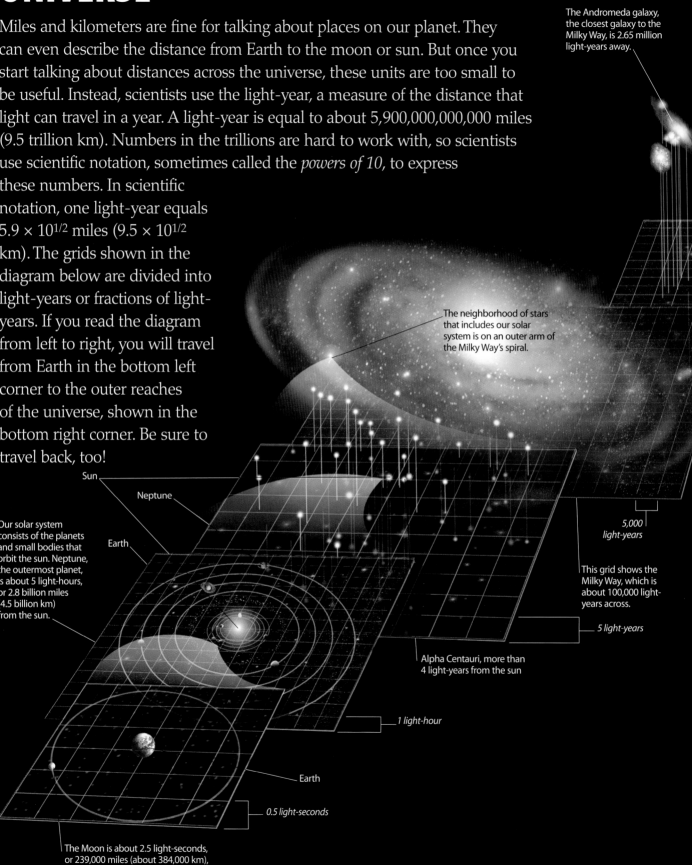

Miles and kilometers are fine for talking about places on our planet. They can even describe the distance from Earth to the moon or sun. But once you start talking about distances across the universe, these units are too small to be useful. Instead, scientists use the light-year, a measure of the distance that light can travel in a year. A light-year is equal to about 5,900,000,000,000 miles (9.5 trillion km). Numbers in the trillions are hard to work with, so scientists use scientific notation, sometimes called the *powers of 10*, to express these numbers. In scientific notation, one light-year equals $5.9 \times 10^{1/2}$ miles ($9.5 \times 10^{1/2}$ km). The grids shown in the diagram below are divided into light-years or fractions of light-years. If you read the diagram from left to right, you will travel from Earth in the bottom left corner to the outer reaches of the universe, shown in the bottom right corner. Be sure to travel back, too!

The Andromeda galaxy, the closest galaxy to the Milky Way, is 2.65 million light-years away.

The neighborhood of stars that includes our solar system is on an outer arm of the Milky Way's spiral.

Sun

Neptune

Earth

Our solar system consists of the planets and small bodies that orbit the sun. Neptune, the outermost planet, is about 5 light-hours, or 2.8 billion miles (4.5 billion km) from the sun.

5,000 light-years

This grid shows the Milky Way, which is about 100,000 light-years across.

5 light-years

Alpha Centauri, more than 4 light-years from the sun

1 light-hour

Earth

0.5 light-seconds

The Moon is about 2.5 light-seconds, or 239,000 miles (about 384,000 km), from Earth.

did you know? ECHOES OF THE UNIVERSE'S BIRTH DURING THE BIG BANG, CALLED *COSMIC BACKGROUND RADIATION*, ADD TO TV STATIC.

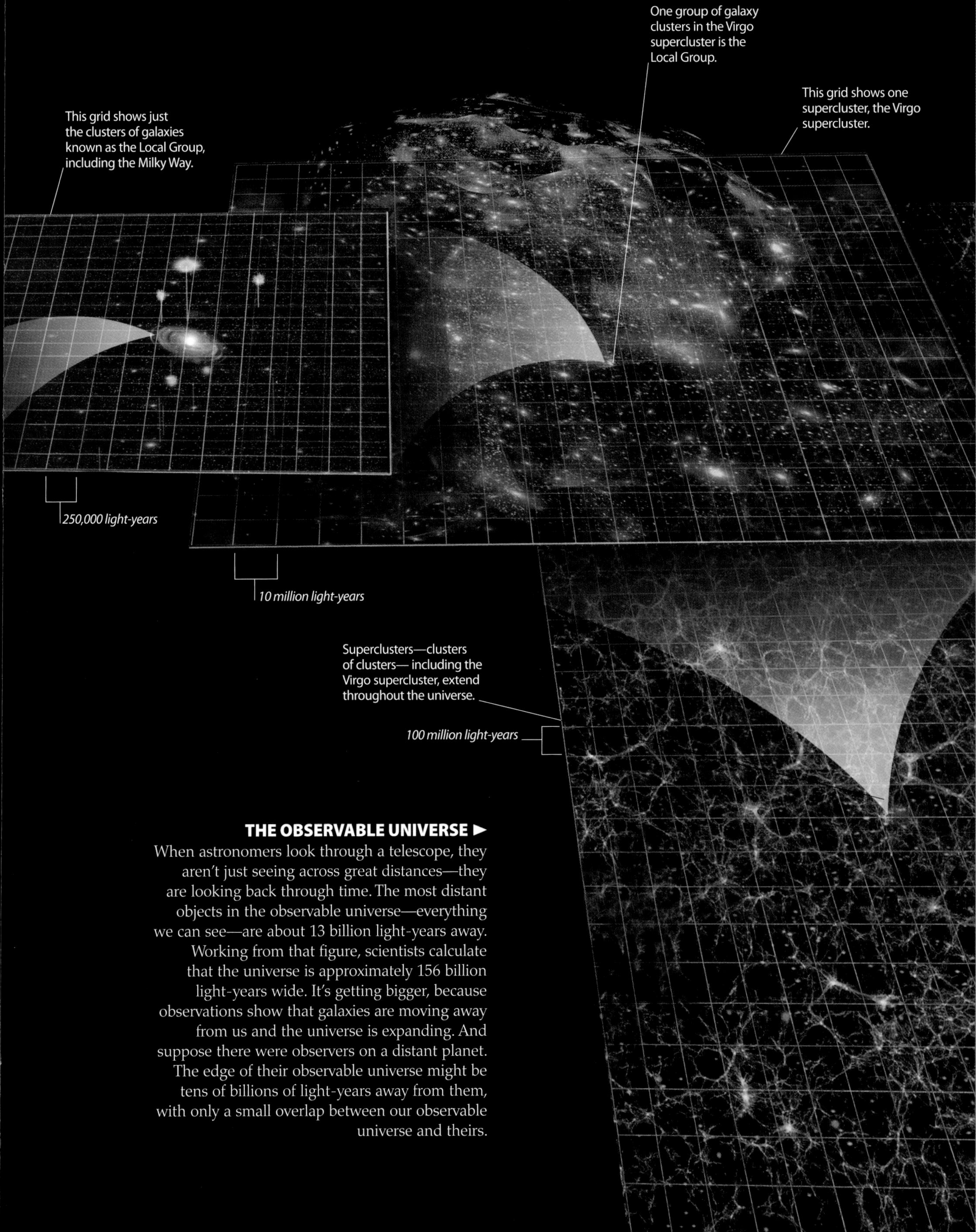

One group of galaxy clusters in the Virgo supercluster is the Local Group.
This grid shows one supercluster, the Virgo supercluster.
This grid shows just the clusters of galaxies known as the Local Group, including the Milky Way.
250,000 light-years
10 million light-years
Superclusters—clusters of clusters— including the Virgo supercluster, extend throughout the universe.
100 million light-years

THE OBSERVABLE UNIVERSE ►
When astronomers look through a telescope, they aren't just seeing across great distances—they are looking back through time. The most distant objects in the observable universe—everything we can see—are about 13 billion light-years away. Working from that figure, scientists calculate that the universe is approximately 156 billion light-years wide. It's getting bigger, because observations show that galaxies are moving away from us and the universe is expanding. And suppose there were observers on a distant planet. The edge of their observable universe might be tens of billions of light-years away from them, with only a small overlap between our observable universe and theirs.

EARTH

It's the only home you know, but is it the only home out there? Earth has many things in common with the other planets in our solar system. It is a sphere. It follows an oval path around the sun. It rotates on an axis, which causes day and night. However, in one important way, Earth is unique. It is the only planet that we know of that is home to living things. No other planet has the conditions needed for life. Earth is near enough to the sun to keep living things warm, but not so near that living things are cooked by its heat. Earth is the only planet whose surface has lots of liquid water, which is necessary for all life as we know it. Surrounding Earth is a blanket of gases called the *atmosphere*. Nitrogen, oxygen, and carbon dioxide are the atmospheric gases that provide living things with the tools they need to harvest energy from sunlight and other materials on Earth's surface.

THE WATERY PLANET ▶

Water exists in three forms, or states, on Earth and in the atmosphere: solid, liquid, and gas. You can see evidence of all three in this picture. The blue oceans are water in its liquid state. The mountains are covered with water in its solid state—ice—making mountaintops snowy-white. When water becomes a gas—water vapor—it returns to the atmosphere, forming clouds.

Clouds of tiny water droplets swirl as if they were wisps of cotton in the atmosphere. When clouds build up and become saturated, the drops fall to Earth as rain.

THE SOLAR SYSTEM ▼

Eight planets orbit the sun. Mercury, Venus, Earth, and Mars are rocky, and Jupiter, Saturn, Uranus, and Neptune are giant balls of gas. Moons, dwarf planets, asteroids, meteoroids, and comets also inhabit the solar system. Moons are balls of rock that orbit planets, while asteroids are chunks of rock that orbit the sun. Comets are chunks of dust and ice that orbit the sun. Meteoroids are pieces that break off from asteroids and comets.

Scientists call Pluto a dwarf planet. Its orbit is at an angle to the orbits of Earth and the other planets.

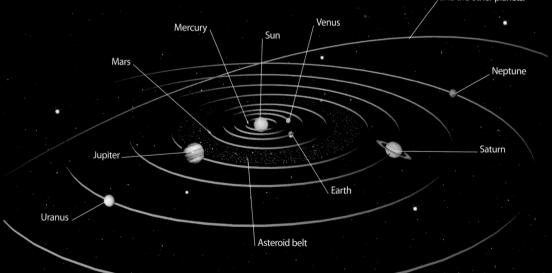

Mercury
Sun
Venus
Mars
Neptune
Jupiter
Saturn
Earth
Uranus
Asteroid belt

did you **know?**.........................
GEOSCIENTISTS CALCULATE THAT THE TEMPERATURE AT THE CENTER OF EARTH'S CORE IS BETWEEN 8,000°F AND 10,000°F (ABOUT 4,427°C–5,538°C)!

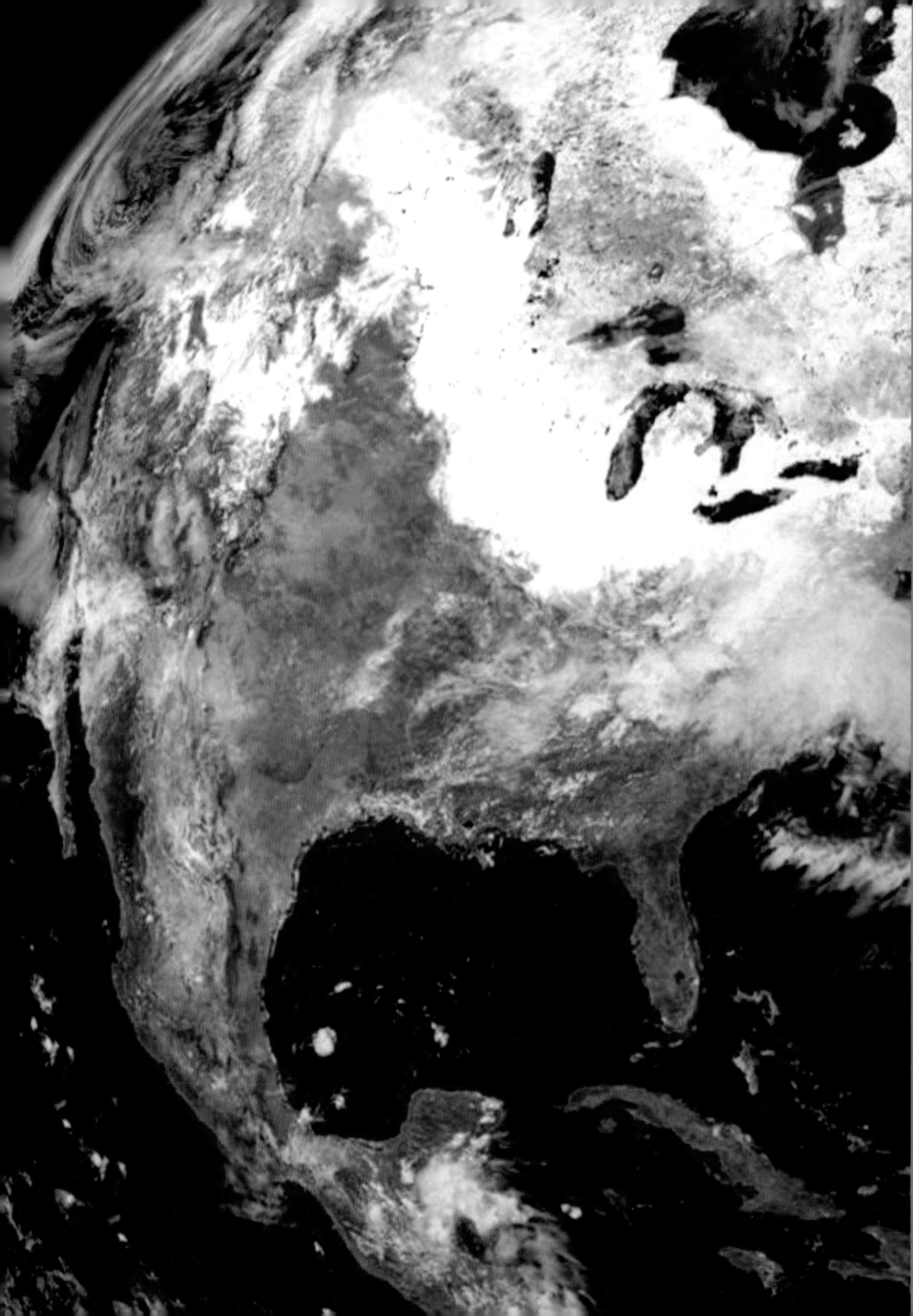

EARTH'S CORE

Science fiction stories tell of journeys to Earth's core, but you really don't want to go there. We live on top of a thin, cool crust. Beneath it are layers that get hotter and hotter the closer you get to the center of Earth. To get to the center, you would have to travel about 3,958 miles (about 6,370 km). It's not too far, really—about the flying distance between Miami, Florida, and Anchorage, Alaska. There in the center, you would find a solid ball of iron and nickel—the inner core. The temperature would be close to 10,000°F (about 5,500°C). Surrounding the inner core is a layer of molten metals—the outer core. Here hot liquid metals rise, cool off, and sink, creating convection currents. Heat from the core causes similar currents in the layer between the core and the crust, called the *mantle*. It is rock, but under such intense pressure and heat, the rock can flow like a slow-moving liquid.

▼ EARTH'S SPACE SHIELD

Geophysicists believe that the currents within Earth's outer core control Earth's magnetic field, called the *magnetosphere*. The magnetosphere protects Earth by deflecting particles. A constant stream of charged particles blows outward from the sun at more than 1 million miles per hour. This solar wind would be deadly if we did not have the protection of a giant magnetic field.

Hot magma in the upper mantle is less dense than the rock around it, so it rises into any cracks in the rock above it.

The magnetic field stretches on the side away from the sun.

The sun's particles are deflected away from the magnetosphere.

The sun produces hot gas that carries particles toward Earth.

The crust is the thinnest layer.

Earth

Solar winds radiate from the sun.

Solar winds squash the magnetic field on the sun side.

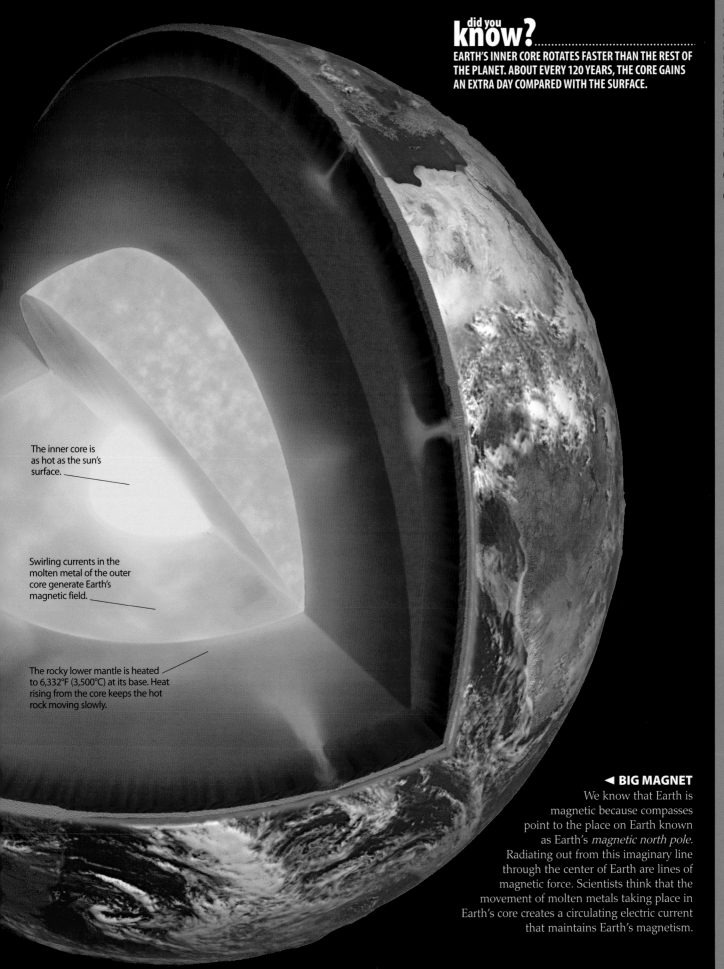

did you know?

EARTH'S INNER CORE ROTATES FASTER THAN THE REST OF THE PLANET. ABOUT EVERY 120 YEARS, THE CORE GAINS AN EXTRA DAY COMPARED WITH THE SURFACE.

The inner core is as hot as the sun's surface.

Swirling currents in the molten metal of the outer core generate Earth's magnetic field.

The rocky lower mantle is heated to 6,332°F (3,500°C) at its base. Heat rising from the core keeps the hot rock moving slowly.

◀ BIG MAGNET

We know that Earth is magnetic because compasses point to the place on Earth known as Earth's *magnetic north pole*. Radiating out from this imaginary line through the center of Earth are lines of magnetic force. Scientists think that the movement of molten metals taking place in Earth's core creates a circulating electric current that maintains Earth's magnetism.

MOON

Just like most of the other planets, Earth has a sidekick—the moon. A moon is a natural satellite, an object that orbits a planet. Earth is nearly 4 times as wide and about 81 times as heavy as its moon. The moon may not be very big, but it has a big effect on Earth. It reflects the sun's light as moonlight during most nights each month. It also affects the Earth's oceans. The moon's gravity, along with that of the sun, causes tides by pulling on Earth's ocean water, causing it to rise up. Throughout history, telescopes and spacecraft have let scientists study the moon from Earth. In the last century, 12 people actually got to visit the moon. Between July of 1969 and December of 1972, American astronauts made 6 successful landings on the moon. On these Apollo missions, astronauts explored the moon's surface, took photographs, collected rock and dust samples, and set up equipment to monitor moon conditions.

◄ SCARRED SURFACE
The moon is covered in craters. Craters are the round dents that form when meteors impact, or hit, the moon's dusty surface. The moon has no atmosphere. This means there is no wind or weather to erode these features after they form. The only things that change craters are geologic activity and newer impacts. Scientists can study the craters to figure out the order in which different moon features formed.

The lunar rover, called a "moon buggy" by some, was a small electric car that let astronauts explore the moon.

During a moon landing, the lunar module would carry two astronauts from a spacecraft in orbit around the moon to the moon's surface and back.

In April of 1972, astronaut John Young jumps up as he salutes the American flag during the fifth moon landing.

▲ APOLLO 16 MOON LANDING

The force of gravity is related to how much an object weighs. Because the moon weighs much less than Earth, its gravity is only one sixth that of Earth. Astronauts visiting the moon during the Apollo missions experienced firsthand its reduced gravity. They found it easier to jump and bounce, even while wearing very heavy spacesuits.

did you know?..

WITH NO WIND, WEATHER, OR LIVING THINGS TO DISTURB THEM, THE EQUIPMENT, DISCARDED VEHICLES, AND EVEN FOOTPRINTS LEFT BY ASTRONAUTS MORE THAN 40 YEARS AGO REMAIN UNCHANGED ON THE MOON'S SURFACE.

▼ MOON OVER EARTH

This photo taken from the International Space Station makes it look as if the moon is in Earth's atmosphere, but it is really only a trick of the light. The moon is actually 238,900 miles (about 384,500 km) away from Earth. During visits to the moon, astronauts have left mirrors there. By bouncing laser light from Earth off these mirrors, scientists can measure exactly how far away the moon is from Earth's surface.

Earth's moon

Earth

SOLAR ECLIPSE

Have you ever observed a solar eclipse? A solar eclipse occurs when the moon passes in front of the sun and casts a shadow on Earth. Imagine the sky getting dark on an otherwise sunny afternoon. For several minutes, it is dark enough to see some of the brightest stars and planets shining in the sky. Moments later, the sky begins to brighten. Within the next hour, it is broad daylight again, as if the brief period of darkness never happened. Because the sun's rays can damage your eyes, you cannot safely view a solar eclipse without specially designed eye protection. With this protection, you can see the dark disk of the moon slowly move in front of the sun, sometimes completely covering it. After several minutes, you can see the moon slowly move past the sun.

▼ THE MOON'S PATH

Several factors determine how long an eclipse will take. The orbits of the moon and Earth and their distances from both each other and the sun are key. Partial phases, during which the moon gradually covers and uncovers the sun, can each take about an hour. A total eclipse can last up to about 7 minutes. This image combines many photos of the stages of the eclipse into one photo. It shows the path of the moon as it passes in front of the sun, as seen from Zambia in Southern Africa.

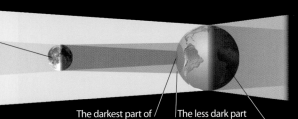

The sun illuminates the side of the moon that we cannot see.

The darkest part of the moon's shadow is called the *umbra*. Where the umbra falls, a total eclipse is visible from Earth.

The less dark part of the moon's shadow is called the *penumbra*. Where the penumbra falls, a partial eclipse is visible from Earth.

Earth

◄ AN ECLIPSE IS POSSIBLE WHEN . . .

A solar eclipse can happen only during the new moon phase, when the moon passes between the sun and Earth and the three are in near-perfect alignment. We don't usually see a solar eclipse, because the moon's path is generally above or below the position of the sun. For a solar eclipse to occur, the moon's path must cross directly in front of the sun.

know?
did you
....
THE NEXT TOTAL SOLAR ECLIPSE THAT WILL BE VISIBLE FROM THE UNITED STATES WILL TAKE PLACE ON AUGUST 21, 2017.

◄ SHOWTIME

Normally the corona—the sun's extended outer atmosphere—is not visible, because the sun's surface, or photosphere, is so bright. The corona can be seen only during a total eclipse, when the moon covers the sun completely. The corona glows around the edges of the moon and can reach temperatures of up to 3.5 million degrees Fahrenheit (2 million degrees Celsius).

The corona extends more than 621,370 miles (1 million km) from the sun's surface. The solar wind originates here.

During a total solar eclipse, prominences—arcs of flaming gases erupting from the sun's surface—can be seen flaring out from behind the moon.

▲ THE PATH OF TOTALITY

Total solar eclipses happen once every year and a half, but the chances of seeing one are slim. The umbra moves across Earth's surface. Its path, called the *path of totality*, is about 167 miles (269 km) in diameter. The shadow travels at about 1,242 miles (2,000 km) per hour—and often crosses the ocean and remote places.

MERCURY

If you could stand on the planet Mercury and look up at the sky, the sun would appear almost three times larger than it does on Earth. However, you would need a specially made spacesuit to protect you. Mercury is so close to the sun that it receives much more heat, light, and dangerous radiation than other planets in our solar system. Its surface temperatures are as high as 806°F (almost 430°C) when the sun is shining and as low as –274°F (almost –170°C) at night. These extraordinary temperatures occur because Mercury has almost no atmosphere. There are not enough gases to trap the sun's heat near the planet's surface at night or to shield it during the day, the way Earth's atmosphere does. Current evidence even suggests that Mercury, the planet closest to the sun, has icy poles. The deep craters at the poles are in permanent shade. Thus, water at the bottom of these craters is always frozen.

did you know?
MERCURY ZIPS AROUND THE SUN IN 88 EARTH DAYS, BUT IT ROTATES SO SLOWLY THAT ONE MERCURY DAY IS TWO THIRDS OF ITS YEAR!

Mercury   Earth   Jupiter   Saturn   Uranus   Neptune

Venus   Mars

Sun

This chart shows the order of the planets. Mercury is closest to the sun.

A sunshade shields the delicate instruments from the sun's extreme heat.

A magnetometer can measure the planet's magnetic field and detect magnetic rocks on the planet's surface.

Solar panels will help power the spacecraft and its many instruments.

◄ A CRATERED SURFACE

With no apparent atmosphere surrounding Mercury to create friction, even small meteors fall to the planet without burning up the way they would passing through Earth's atmosphere. Also, there is no wind to erode, or blow away, the marks the meteors make when they land on Mercury's surface. Thus, Mercury looks like Earth's moon—covered with craters, the round depressions that form when objects hit its surface.

▲ MESSENGER

Launched from Earth in 2004, the MESSENGER spacecraft is designed to orbit Mercury starting in 2011. Scientists hope the data it sends back will help them better understand the shapes on Mercury's surface, the composition of the planet's rocks and atmosphere, and how its magnetic field works.

▼ MERCURY'S TRANSIT OF THE SUN

A transit is similar to a solar eclipse. It happens when a planet passes in front of our view of the sun. To see a transit, you need a telescope that filters out most of the sun's light. In this image, Mercury looks like a marble against the massive sun. The most recent transit of Mercury took place in 2006. The next ones happen in 2016 and 2019.

Hot gas surrounds the sun at its surface.

Mercury crosses the face of the sun.

VENUS

Venus may seem similar to Earth in size and mass, but don't plan to spend your next vacation there. Its poisonous clouds of sulfuric acid, metal-melting heat, crushing atmospheric pressure (almost 90 times greater than Earth's), hurricane-like winds, and suffocating atmosphere of carbon dioxide make the second planet from the sun an unwelcoming place. The clouds of Venus trap the baking heat near the planet's surface, causing temperatures of more than 870°F (about 466°C). These same clouds also reflect lots of sunlight, making Venus one of the brightest objects that we see in the night sky. Venus has retrograde rotation, meaning it spins in the opposite direction from the direction in which it travels around the sun. This means that the sun rises in the west and sets in the east—the opposite of Earth. However, it spins very slowly, taking 244 Earth days for one complete spin. On the other hand, it takes Venus only about 225 Earth days to circle the sun. So a Venusian day is actually longer than a Venusian year!

VENUS EXPRESS ▶

Because Venus is so hot and its atmospheric pressure so great, studies of the planet are made from unmanned orbiting spacecraft. In 2005, the European Space Agency launched the Venus Express spacecraft to study Venus. Equipped with infrared and ultraviolet cameras to map the surface, the spacecraft has gathered important information about the composition of the atmosphere and the nature of the planet's magnetic field. Venus Express will send information to scientists on Earth through December 2012.

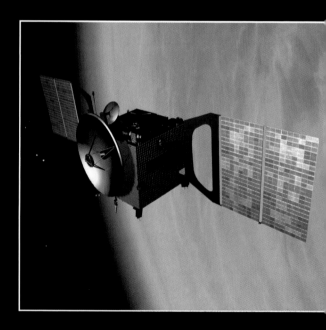

did you know?

ALTHOUGH IT IS ALMOST TWICE AS FAR FROM THE SUN AS MERCURY, VENUS IS THE HOTTEST PLANET IN THE SOLAR SYSTEM. ITS THICK ATMOSPHERE TRAPS HEAT NEAR THE SURFACE.

Solar System — Mercury — Earth — Jupiter — Saturn — Uranus — Neptune — Venus — Mars — Sun

During its orbit around the sun, Venus comes closer to Earth than any other planet, about 23.7 million miles (38.2 million km).

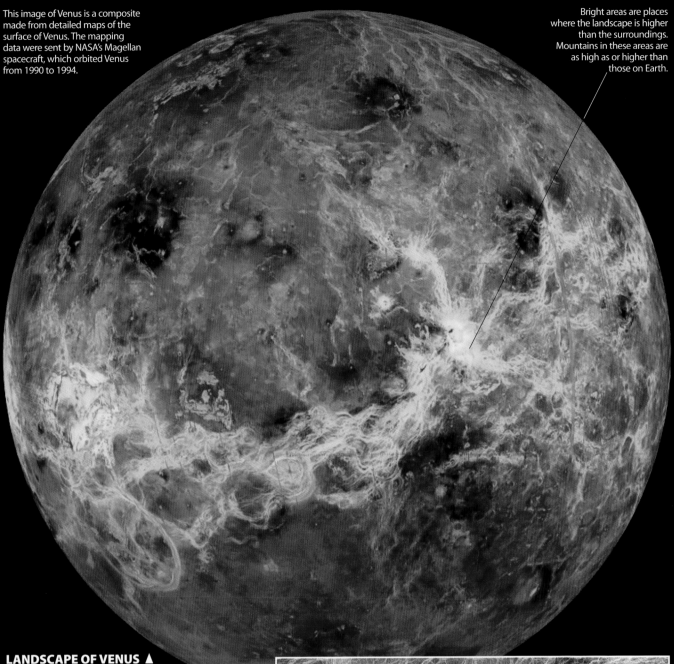

This image of Venus is a composite made from detailed maps of the surface of Venus. The mapping data were sent by NASA's Magellan spacecraft, which orbited Venus from 1990 to 1994.

Bright areas are places where the landscape is higher than the surroundings. Mountains in these areas are as high as or higher than those on Earth.

LANDSCAPE OF VENUS ▲

Venus's landscape is hard to see from an orbiting satellite because of its thick atmosphere. Using radar and other electromagnetic wave technology, scientists have discovered that Venus has mountains, valleys, canyons, and volcanoes. One mountain range is about 7 miles (about 11 km) high. That's almost 1.5 miles (about 2 km) higher than Earth's highest peak, Mount Everest!

MYSTERIOUS MARKINGS ▶

Giant spiders on Venus? Not really! These markings on the surface of Venus resemble spider webs and are called *arachnoids*—a term used to classify spiders. Each is about 30 to 140 miles (about 48 to 225 km) wide. Scientists don't know exactly how they formed but they think the markings came about when melted rock called *magma* pushed up from underground, causing cracks in the surface rocks.

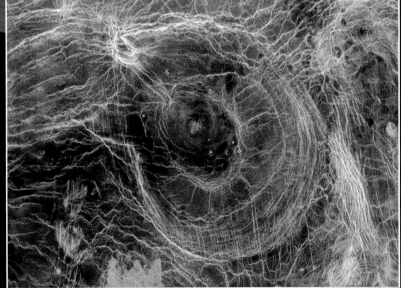

267

MARS

Mars has always captured people's imaginations. In the nineteenth century, an Italian astronomer described what he called "canalis" on the surface of Mars. Later, an American scientist translated the term as "canals," implying that they had been made by "intelligent beings." Although some people still believe Martians exist, none of the missions to Mars has returned any evidence to support this theory. In truth, Mars is a small planet whose surface has been shaped by volcanoes, quakes, dust storms, and impacts from meteors. It is very cold, on average between –135°F and 26°F (about –93°C and –3°C). Its atmosphere is about 95 percent carbon dioxide, making it unsuitable for human life. Mars exploration began in the 1960s and continues to this day using unpiloted machines called *space probes.* In 2008, NASA's Mars lander, Phoenix, reached the surface of the planet and transmitted information about its climate and geology to Earth for 5 months. Phoenix confirmed the presence of water ice on Mars, and not just a little ice. There is enough frozen water to fill Lake Michigan twice!

did you know? THE LENGTH OF ONE MARTIAN YEAR IS APPROXIMATELY EQUAL TO 687 EARTH DAYS.

GIANT DUST STORMS ▶

Huge dust storms have been known to envelop Mars in a matter of weeks. For example, in 2001, Hubble Space Telescope images showed the beginning of one such dust storm. For more than 3 months, scientists observed as the dust storm grew until it blanketed the whole planet. During this time, the air temperature on Mars rose 54°F (30°C).

Mars is often called the Red Planet because its soil has iron oxide in it.

Solar System

Mercury Earth Jupiter Saturn Uranus Neptune

Venus Mars

Sun

Mars is the fourth planet from the Sun. This image shows the order of the planets and their approximate sizes relative to one another.

Olympus Mons is about as wide as the state of New Mexico, and is about 15 miles (25 km) tall. That's almost three times as tall as Mt. Everest!

▲ THE BIGGEST VOLCANO

Mars is about half the size of Earth, but it has perhaps the largest volcano in our solar system—Olympus Mons. Most volcanoes on Mars are 10 to 100 times larger than those on Earth.

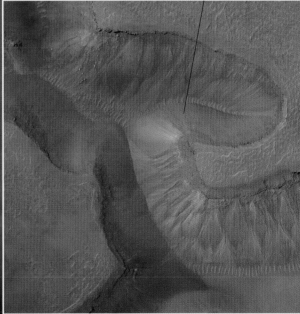

Gullies possibly made by flowing water

▲ IS THERE LIQUID WATER ON MARS?

The Martian temperature is too cold for liquid water to exist on the surface. However, scientists think that liquid water may have flowed on the surface of Mars hundreds of billions of years ago when the planet was warmer than it is today.

Polar ice cap

JUPITER'S MOONS

Jupiter, the largest planet in our solar system, has at least 64 known moons. That is more than any other planet! All but four moons are very tiny—not even a hundredth of the size of Earth's moon. Most of the moons are simply referred to as Jupiter's natural satellites. More than half of them travel in the opposite, or retrograde, direction of the planet's spin. Astronomers believe this motion indicates that these satellites were once asteroids fragmented by collisions and captured by Jupiter's tremendous pull long after its four larger moons were formed. Today, we know more about Jupiter's moons thanks to the information obtained by a NASA spacecraft called *Galileo*. Equipped with many scientific instruments, it orbited Jupiter between 1995 and 2003. During that time, *Galileo* transmitted to Earth thousands of images and readings of the planet and its four largest moons: Ganymede, Callisto, Io, and Europa.

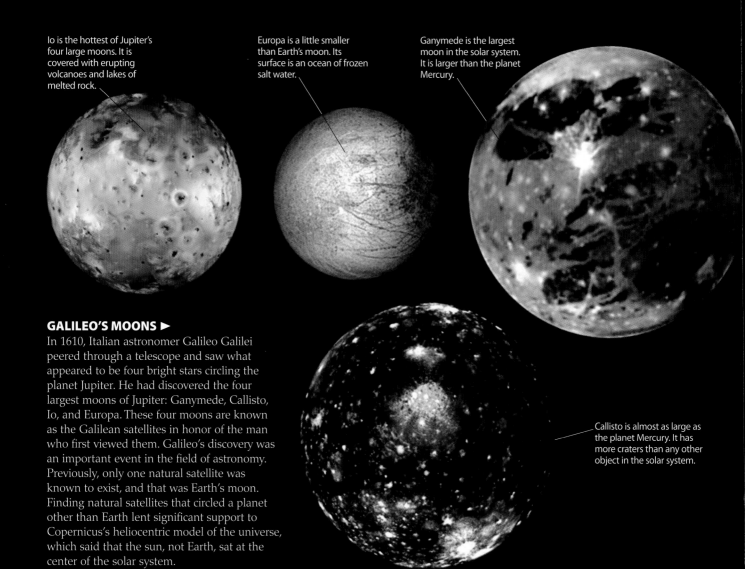

Io is the hottest of Jupiter's four large moons. It is covered with erupting volcanoes and lakes of melted rock.

Europa is a little smaller than Earth's moon. Its surface is an ocean of frozen salt water.

Ganymede is the largest moon in the solar system. It is larger than the planet Mercury.

Callisto is almost as large as the planet Mercury. It has more craters than any other object in the solar system.

GALILEO'S MOONS ▶

In 1610, Italian astronomer Galileo Galilei peered through a telescope and saw what appeared to be four bright stars circling the planet Jupiter. He had discovered the four largest moons of Jupiter: Ganymede, Callisto, Io, and Europa. These four moons are known as the Galilean satellites in honor of the man who first viewed them. Galileo's discovery was an important event in the field of astronomy. Previously, only one natural satellite was known to exist, and that was Earth's moon. Finding natural satellites that circled a planet other than Earth lent significant support to Copernicus's heliocentric model of the universe, which said that the sun, not Earth, sat at the center of the solar system.

Mercury Earth Jupiter Saturn Uranus Neptune

Venus Mars

Sun

Jupiter is the fifth planet from the sun. This image shows the order of the planets and their approximate size relative to one another.

Beneath Jupiter's stormy atmosphere, the pressure is so great that hydrogen gas is compressed to a liquid state.

Ganymede is the solar system's largest moon. It is larger than the planet Mercury.

▲ JUPITER THE GIANT

Jupiter is a giant ball with a gaseous surface, not a solid surface like Earth's. It is so huge that about 1,300 Earths would fill its volume! Jupiter appears to be covered with stripes of different colors. These alternating dark belts and light zones are created by strong winds in the planet's upper atmosphere. Storms rage within these belts and zones.

did you know? IO'S VOLCANIC ACTIVITY IS 100 TIMES GREATER THAN EARTH'S! IT IS THE MOST VOLCANICALLY ACTIVE BODY IN OUR SOLAR SYSTEM.

SATURN

About 700 B.C., the ancient Assyrians thought Saturn was a very brilliant star. For several centuries, people thought the object was a wandering star. In 1610, Italian astronomer Galileo Galilei viewed Saturn through a telescope. He was the first person to see Saturn's rings. The second-largest planet in our solar system, Saturn has a diameter of about 74,900 miles (about 120,540 km). That is almost ten times Earth's diameter. However, the planet is not solid like Earth. It is a large ball of gas with a solid inner core that is very hot—nearly 21,140°F (about 11,730°C). And, Saturn's atmosphere is very cold—close to −288°F (about −178°C). Saturn also has at least 63 moons. Some are very tiny, like Aegaeon, which measures about 0.3 miles (about 0.5 km). Others are huge, like Titan, which is bigger than the planet Mercury!

THE RINGS ▶

Saturn's most spectacular feature is its rings. These are made of billions of pieces of ice and rock that circle around the planet at different speeds and span about 175,000 miles (almost 282,000 km). Astronomers have now identified seven sets of rings, although not all are visible in this photo. Each set of rings is composed of many ringlets. Astronomers also think that these rings are the remnants of comets, asteroids, and moons that broke up and were captured by Saturn's gravitational pull.

did you know?...

SATURN IS ONE OF THE WINDIEST PLANETS IN OUR SOLAR SYSTEM. WINDS THERE CAN REACH 1,100 MILES PER HOUR (ABOUT 1,800 KM/H)!

Solar System

Mercury   Earth   Jupiter   Saturn   Uranus   Neptune

Venus   Mars

Sun

Saturn takes 29 Earth years to complete one orbit around the sun.

MOONS AND PROBES ▶

Scientists continue discovering more moons orbiting Saturn. Before NASA's Cassini-Huygens space probe was sent to Saturn in 1997, only 18 moons orbiting the planet had been identified, as shown here. Today, more than 63 have been discovered! NASA has sent four space probes to Saturn: Pioneer 11, Voyager 1, Voyager 2, and Cassini-Huygens, which arrived in 2004 and is still sending data back to Earth.

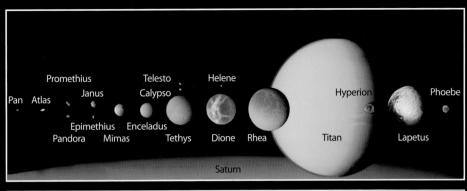

Pan Atlas Promethius Janus Telesto Calypso Helene Hyperion Phoebe
Epimethius Enceladus Dione Rhea Titan Lapetus
Pandora Mimas Tethys
Saturn

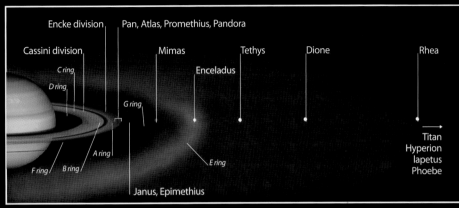

Encke division Pan, Atlas, Promethius, Pandora
Cassini division Mimas Tethys Dione Rhea
C ring Enceladus
D ring
G ring
A ring Titan
F ring B ring E ring Hyperion
Iapetus
Janus, Epimethius Phoebe

The Cassini division is a space that separates Saturn's B and A rings.

Saturn's seven rings are named with the letters A through G in the order they were discovered. Looking outward from the planet, the rings are D, C, B, A, F, G, and E.

Saturn spins on its axis much faster than Earth does. A day on Saturn averages 10 Earth hours, 14 minutes.

URANUS

Imagine the distance from the sun to Saturn: 941,070,000 miles (1,514,505,358 km) at its farthest point. Now double it. That's about where you'll find the seventh planet from the sun, Uranus. Like Jupiter, Saturn, and Neptune, Uranus is a giant ball of gas and liquid, primarily hydrogen and helium, with small amounts of ammonia, water, and methane ice crystals. Beneath the visible clouds is a layer of liquid—under exteme pressure—made up mostly of water, ammonia, and methane. At the center of the planet, scientists think there may be a rocky core about the size of Earth. Like all of the other planets, Uranus makes an oval-shaped orbit around the sun, which it completes in 30,685 Earth days (84 Earth years). What makes this planet different from all the other planets in our solar system? It spins almost on its side, like a slightly tilted Ferris wheel. Scientists hypothesize that, soon after it formed, an Earth-sized object struck Uranus, pushing it over.

Umbriel is a very old moon that has many large craters on its unusually dark surface. Astronomers do not know why Umbriel is so dark.

A layer of methane gas covers the clouds in the outer atmosphere of Uranus, giving the planet its blue-green color.

Ariel is Uranus's brightest moon, with perhaps the youngest surface. Its many valleys are marked with numerous small craters.

did you know? IF THE SUN STOPPED SHINING, IT WOULD TAKE OVER 2.5 HOURS FOR URANUS TO BE IN DARKNESS.

▼ MANY MOONS

Scientists have so far discovered 27 moons of Uranus—most of them named after characters from the plays of William Shakespeare. Uranus's five largest moons, the first two of which were discovered in 1787, are each less than 1,000 miles (1,600 km) in diameter. Ten of its smaller moons were first identified from pictures taken by the Voyager 2 spacecraft during its flyby of Uranus in 1985 and 1986. The smallest known moons circling Uranus are only 8–10 miles (12–16 km) across.

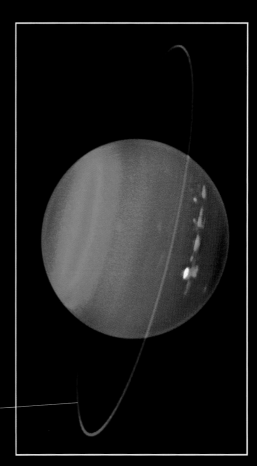

Titania is Uranus's largest moon and one of the first two moons discovered.

Like Umbriel, Oberon is covered with large craters. It contains equal parts of rock and ice, similar to Uranus's other large moons.

Uranus ring system

▲ RINGS AROUND THE PLANET

In 1977, scientists first observed the presence of the rings of Uranus when the planet passed in front of a star. With the help of Voyager 2 and the Hubble Space Telescope, scientists have identified 13 rings so far. One of the last 2 rings to be discovered is about 60,708 miles (97,700 km) from the planet's center.

The surface of Miranda is an unusual jumble of features, including both parallel and crisscrossing canyons and ridges.

Solar System

Mercury   Earth   Jupiter   Saturn   Uranus   Neptune

Venus   Mars

Sun

Uranus is one of the Jovian, or Jupiter-like, planets. It has the third-largest diameter of all the planets in the solar system.

NEPTUNE

In the early 1800s, astronomers knew of only 7 planets in our solar system. Although Galileo had seen a bright "star" through his telescope in 1613, he didn't realize that it was a planet. In 1845, the French mathematician Urbain Le Verrier realized that the orbit of Uranus—the seventh known planet— was very different from what his calculations predicted it should be. He reasoned that the only way Uranus could move the way it did along its orbit was if another large planet's gravity was pulling on it. In 1846, German astronomer Johann Gottfried Galle used Le Verrier's prediction and found that large planet was Neptune! Neptune is about 4 times as wide as Earth and 17 times more massive.

Planet Neptune is one of the four gas giants in our solar system. Its solid core is about the size of Earth. Surrounding its icy, rocky core is a layer of liquids, including water, and a thick, cloudy atmosphere.

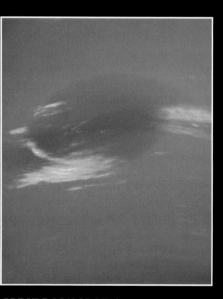

A STORMY SURFACE ▶

Neptune's thick layer of gases and clouds has winds that blow faster than 1,000 miles per hour (about 1,600 km/h). The clouds on the surface are made up of methane—the same gas you burn in a gas stove. Methane gives the planet its blue color. Darker clouds that lie beneath the methane are probably made of hydrogen sulfide—the chemical that gives rotten eggs their yucky smell.

GREAT DARK SPOT ▲

In 1989, an NASA spacecraft named *Voyager 2* took pictures of a dark storm on Neptune's surface. This swirling, hurricane-like storm was almost the size of Earth! Scientists called it the Great Dark Spot. Five years later, it was nowhere to be found, but another huge storm, seen in 1994, lasted 3 years. Imagine what would happen to Earth if a hurricane lasted for 3 years!

Solar System

Mercury
Earth
Venus
Mars
Sun
Jupiter
Saturn
Uranus
Neptune

This chart shows the order of the planets from the sun. Neptune is the eighth planet from the sun.

At an average 2.8 billion miles (about 4.5 billion km) from the sun, Neptune takes 165 Earth years to make 1 complete revolution around the sun.

▲ NOT A CANTALOUPE ...A MOON!

Neptune has 13 known natural satellites, or moons, of which the moon Triton is the largest. Its surface resembles a cantaloupe and has temperatures of about –400°F (–240°C). Triton is one of the coldest places in the solar system! It is so cold that, instead of lava, volcanoes spew icy mixtures that freeze like snow as they fall back to the ground. Triton also orbits Neptune in the opposite direction of the planet's rotation, leading scientists to believe that this moon was captured by the planet's gravity long after the two had formed.

did you know?...
NEPTUNE IS SO FAR AWAY FROM EARTH THAT IT IS THE ONLY PLANET IN OUR SOLAR SYSTEM THAT CANNOT BE SEEN WITHOUT A TELESCOPE.

From the time of its discovery in 1930, Pluto was the ninth planet in our solar system. But science changes, and some things are hard to let go. That's what happened in 2006 when astronomers decided to stop classifying Pluto as a planet. Why? A celestial body must meet three characteristics to be called a planet: it must orbit the sun, it must be round, and it must have cleared away any other objects in its orbit. Pluto comes close, but it has other objects occupying its orbit. Therefore, scientists now call Pluto a dwarf planet. Pluto also has three small moons named Charon, Hydra, and Nix. The space probe New Horizons, sent by NASA in 2006 to explore Pluto and its moons, will arrive in 2015. It's a long trip because the dwarf planet is about 3.6 billion miles (about 5.8 billion km) away from the sun. That's about 39 times farther than Earth is from the sun!

The distant sun would look much dimmer from Pluto than it would from Earth.

Scientists think that Charon's surface is covered in frozen water, while Pluto's surface is probably a frozen mixture of methane, nitrogen, and carbon monoxide.

PLUTO FROM CHARON

The background scene shows how the sun and Pluto would look if you were standing on the surface of Charon (below). Just as our moon shows only one side to Earth, Charon shows only one side to Pluto. Pluto also rotates so that it shows only one side to Charon at all times! This is called *tidal locking*. It happens because Charon's orbit around Pluto takes about 6.5 Earth days, and one full rotation of Pluto also takes about 6.5 Earth days.

Solar System

Mercury   Earth   Jupiter   Saturn   Uranus   Neptune

Venus   Mars   Pluto

Sun

The dwarf planet Pluto is compared with the 8 planets of our solar system.

When Pluto is farthest from the sun, its atmosphere freezes and falls to the ground.

HOW BIG IS PLUTO? ▶

When you look at a planet, you probably think of something that is much larger than Earth's moon. Pluto's diameter is about 1,441 miles (2,320 km), making it smaller than Earth's moon. Charon, Pluto's largest moon, has a diameter of about 790 miles (about 1, 270 km), about half the size of Pluto. Farther away than Pluto, astronomers have discovered two small objects that orbit our sun at the very fringe of the solar system. They have named them *Sedna*, after the Inuit goddess of sea creatures, and *Quaoar*, the god of creation of the Tongva people of California.

Sedna
(800–1,100 miles
in diameter)

Quaoar
(800 miles)

Pluto
(1,400 miles)

Moon
(2,100 miles)

Earth
(8,000 miles)

ASTEROIDS

If you can imagine a rock the size of a city moving through space, you have a good idea of what an asteroid is. Asteroids are rocky objects that orbit the sun. Some are smaller than a house, but the largest is almost as wide as Texas. Most asteroids are in the asteroid belt between the orbits of Mars and Jupiter. The millions of asteroids there are probably left over from when the solar system formed. Some asteroids have orbits that extend past Neptune, the farthest planet. Others have orbits that bring them close to Earth. Astronomers watch for these asteroids because of the damage they could cause if they were to hit Earth. Fortunately, asteroids large enough to cause widespread destruction hit Earth only every thousand years or so.

▼ **WHAT'S MISSING?**
Notice something missing in this picture? Unlike most asteroids, which are covered with craters from their many collisions, Itokawa has no craters. No one knows exactly why, but one idea is that Itokawa's rocks are so loosely held together that collisions shake piles of rubble into the craters, filling in the holes.

▼ **EXPLODING FAMILIES**
Families aren't just for people. Most asteroids in the asteroid belt are members of families, too! Families are groups of asteroids with similar properties and orbits. The steps below show how an asteroid family forms.

1 A stray asteroid or other object moves through the solar system on a collision course with an asteroid. The asteroid it hits is called the *parent asteroid*.

2 Before the collision, the parent asteroid is in orbit within the asteroid belt.

3 When the object crashes into the parent asteroid, its energy is transferred to the spot where the object hits.

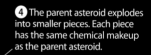

4 The parent asteroid explodes into smaller pieces. Each piece has the same chemical makeup as the parent asteroid.

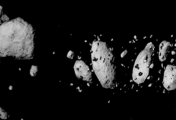

did you know? A MASSIVE ASTEROID THAT CRASHED INTO EARTH MAY HAVE CAUSED THE EXTINCTION OF DINOSAURS.

▲ BAKED IN SPACE?

Ida was named for a Greek nymph, not for its resemblance to an Idaho potato. Ida is about 37 miles (60 km) wide. You can tell from the number of craters that Ida is held together more strongly than Itokawa is. Ida is different from most asteroids because it has a moon! Ida's moon is less than a mile (1.5 km) wide. This moon, Dactyl, was the first ever discovered orbiting an asteroid.

CERES—A ROUND ASTEROID ▼

Have you ever wondered why planets are round? Ceres—the largest asteroid in the asteroid belt—reveals the answer. In order for an object in space to be round, it must have enough mass that its own gravity pulls it into a round shape. Ceres is about 600 miles (960 km) wide. Its mass is great enough to pull it into a ball. Also like the planets, Ceres has layers—a core, a mantle, and a crust.

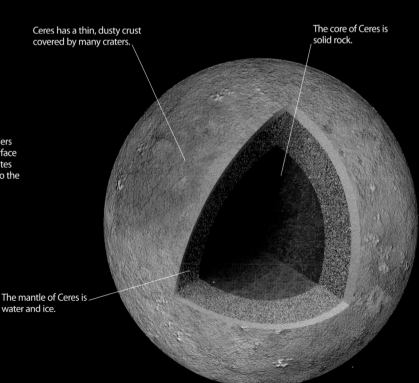

Ceres has a thin, dusty crust covered by many craters.

The core of Ceres is solid rock.

The giant boulders on Itokawa's surface may be meteorites that crashed into the asteroid.

The mantle of Ceres is water and ice.

Over time, the gravitational pulls of Mars and Jupiter cause the pieces of the new asteroid family to settle into orbits that are similar to one another.

METEORITES

On a clear night, a bright streak shoots across the sky. A chunk of metal or rock has just sped into Earth's atmosphere from space. The chunk slammed into the air, generating so much heat from friction that it began to burn, leaving behind a trail of glowing gas called a meteor or a shooting star. The upper atmosphere shatters most meteors into tiny bits, but some large rocks may survive and crash into Earth's surface. The pieces that survive are called meteorites. In all of human history, a meteorite has struck a person only once. In Alabama in 1954, Ann Hodges was on her couch when an 8.5 pound (3.8 kilogram) meteorite broke through her roof. A bit bigger—and heavier—than a brick, it hit her hand and hip, creating large bruises.

did you know?
THE LARGEST METEORITE EVER FOUND IS ABOUT 3 FEET (1 METER) HIGH AND 8 FEET (2.5 METERS) WIDE. IT WEIGHS MORE THAN 60 TONS.

Why is an iron meteorite red? The iron reacted with oxygen after it reached Earth. The red color is from rust.

The crater is 570 feet (170 meters) deep.

▲ **IRON-NICKEL METEORITE FROM NAMIBIA**
This meteorite was once part of the dense center of a large asteroid. Now it is a small, heavy chunk of iron with a tiny amount of nickel. Meteorites can provide direct evidence of what our solar system was like when it began about 4.6 billion years ago.

METEOR ▶

Meteors fly into the atmosphere at speeds as high as 100,000 miles per hour (160,000 kilometers per hour). During a meteor shower, it is sometimes possible to watch 100 meteors bombard our planet each hour. Meteor showers occur when Earth travels through particles left behind by a comet.

The crater rim rises 150 feet (45 meters) above the surrounding plains.

▼ IMPACT CRATER IN ARIZONA, USA

The largest impacts from meteorites gouge craters out of our planet. About 50,000 years ago, an iron-nickel meteorite 150 feet (45 meters) across smashed into northern Arizona. Within seconds, the impact melted all the rock nearby and pushed out the molten rock with enough explosive force to dig out this crater, 4,000 feet (1,200 meters) in diameter. The meteorite itself shattered into tiny bits of metal. These dense fragments sank in the lighter molten rock. Soon the rock above cooled and turned into a solid again. Today, geologists believe small pieces of the original meteorite are trapped deep below the crater floor.

Crystals of the mineral olivine are embedded in iron.

▲ STONY-IRON METEORITE FROM ARGENTINA

The mixture of stone and metal in this meteorite provides geologists with good evidence about its origins. It was once part of the region where an asteroid's core and crust met. The stony parts came from the asteroid's lighter crust and the iron parts from the asteroid's dense core.

MILKY WAY

The Milky Way is a galaxy—a vast group of stars, dust, gas, planets, and asteroids pulled together by the force of gravity. It contains more than 100 billion stars and planets, including the sun and the rest of our solar system. The Milky Way is more than 100,000 light-years long! A light-year is the distance light travels in one year (5.88 trillion miles, or 9.46 trillion km), so the Milky Way is enormous! The Milky Way formed more than 13 billion years ago. To the naked eye, it looks like a broad white band in the night sky. Ancient people thought this white band looked like a river of milk, so they named it the Milky Way.

SPIRAL GALAXY ▶
The Milky Way is a spiral galaxy, a galaxy shaped like a pinwheel. A spiral galaxy's arms coil outward from the center and are regions where stars form. Thousands of hot, young, blue and blue-white stars give the arms a bright appearance. Like all spiral galaxies, the Milky Way slowly rotates—so slowly it takes about 250 million years for our sun to circle the galaxy's center!

Young blue and blue-white stars shine much brighter than older, redder stars.

did you know? OUR SUN IS LOCATED ABOUT 26,000 LIGHT-YEARS AWAY FROM THE CENTER OF THE MILKY WAY!

BARRED SPIRAL GALAXY ▲
Scientists believe there are more than 100 billion galaxies in the universe. The common barred spiral galaxy has a band of bright stars that emerges from the center and extends across the middle of the galaxy. About half of all spiral galaxies are barred. Many scientists believe the Milky Way is a barred spiral galaxy.

THE SOMBRERO GALAXY ▲
From Earth, we can see the Sombrero galaxy only edge-on. Its flat disk and bulging center, formed by billions of old glowing stars, give the galaxy its hatlike appearance. Bands of dust form the disk's dark rings. Scientists do not fully understand them, but believe the rings contain younger and brighter stars. They also believe the center of the galaxy holds a large black hole.

THE MILKY ROAD ▼
This image was created from different photos, to make the Milky Way look as though it extended out from the end of a road. From Earth, the Milky Way does not look this colorful. Also, it is difficult to see, especially in cities where the night sky is filled with light from cars, buildings, and streetlights.

At the center of the Milky Way may be a black hole a million times larger than the sun.

BIG BANG THEORY

We know that the universe is huge . . . and old . . . but how huge and old is it? And, how did it form? Scientists use mathematics to test ideas about just how the universe came to be what it is today. The idea that is accepted by most scientists is called the *Big Bang theory*. The Big Bang theory states that the universe began as an infinitely small point. This point contained all of the matter and energy in the universe today. Suddenly, a huge expansion occurred, called the *Big Bang*. As it expanded, the universe cooled and its matter spread far apart. It is still enlarging today. Scientists have found evidence of the Big Bang theory by studying galaxies. They found that the farther a galaxy is from Earth, the faster it is moving away from Earth. This can be true only if the universe is expanding in all directions.

TIMELINE OF THE UNIVERSE
This model shows how the universe formed 13.7 billion years ago. It shows the development from a tiny point to the 93-billion-light-years-wide observable cosmos that it is today. Keep in mind that the sizes, distances, and times shown here are not to scale. The universe is so large and so old that it would be impossible to fit its history onto the pages of a book.

At the Big Bang, the universe is extremely small, bright, dense, and hot.

Radiation fills the universe.

Less than a second after the Big Bang, the temperature is 10 million trillion degrees Celsius. Simple particles form.

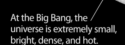

did you
know?..................................
ASTRONOMERS ONCE THOUGHT THAT WHEN THE UNIVERSE WAS DONE EXPANDING, IT WOULD START TO SHRINK. THEY CALL THIS SHRINKING THE "BIG CRUNCH."

Not quite 2 minutes after the Big Bang, the temperature is ten billion

It's 380,000 years after the Big Bang, and the temperature is almost 5,000°F (3,000°C). Clumps of gas form.

One billion years after the Big Bang, the temperature is -450°F (-255°C). The first galaxies begin to form.

Three billion years after the Big Bang, the temperature of space reaches what it is today, about -454°F (-270°C).

About nine billion years after the Big Bang, the sun forms inside the Milky Way galaxy. Earth forms from leftover material at about the same time.

BLACK HOLES

Start with a star 10 times more massive than the sun.
When the star dies, its center, or core, collapses, and its
outer layers fly out in a spectacular explosion called a
supernova. The core continues to shrink, becoming a black
hole that might be only 20 miles (about 32 km) wide and
incredibly dense. Once anything, including a light wave,
enters a black hole, it can never leave. All of the mass of a
black hole is at its center in a point called a *singularity*.
In a sense, the singularity is a hole in the universe
that soaks up matter and energy.

Matter spiraling around the black
hole is called its *accretion disk*.
Until it gets very close, this matter
does not act any differently than
matter around a large star.

As the matter gets very close to
the black hole, it gains energy and
heats to several million degrees.

STAR POWER ▼

Gravity pulls a star's mass toward its core. This inward
pressure balances the outward pressure from the energy
that is created when stars combine lighter elements, such
as helium and carbon, into heavier ones, such as oxygen.
This process of combining elements is called *fusion*.
Big stars can produce enough heat to continue fusing
elements—until they get to iron. Fusing iron with other
elements uses energy rather than creating it, so fusion
stops. Then gravity wins, and the star collapses.

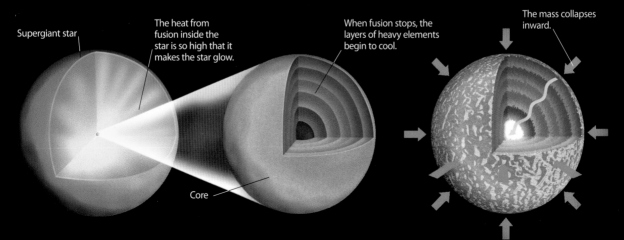

Supergiant star

The heat from
fusion inside the
star is so high that it
makes the star glow.

When fusion stops, the
layers of heavy elements
begin to cool.

The mass collapses
inward.

Core

▼ A COSMIC WHIRLPOOL

If black holes don't let light out, how can we find them? When a black hole is near some other stuff—say, gas or rocks—its gravity gives a tug and the matter begins to spiral toward the black hole. The matter moves faster and faster, picking up energy. High-energy matter sends out high-energy radiation. Back home in the solar system, we use space telescopes to grab photos of the X-rays that come from really, really hot, high-energy matter.

did you know? A BLACK HOLE WITH THE MASS OF EARTH WOULD FIT IN A SPHERE THE SIZE OF A MARBLE.

Some of the matter around the rapidly spinning black hole flies off at speeds near the speed of light. The X-ray energy emitted by these jets of matter is evidence of the black hole.

If you could watch as the astronaut moved toward the hole, she would appear to move more and more slowly. You would never see the astronaut cross the event horizon.

The boundary beyond which light cannot escape is called the *event horizon*. This is the black hole's diameter as viewed from the outside.

At the event horizon, the light takes almost forever to get away from the black hole, so an image of the astronaut remains long after the astronaut has passed.

DOWN THE HOLE ▶

What would you see if you watched someone fall into a black hole? The term "spaghettification" was coined after physicist Stephen Hawking said the astronaut in the illustration resembled a piece of spaghetti being pulled long and thin. The fantastic mass of the black hole exerts a great gravitational force on objects when they are very close. The force is much stronger on the astronaut's head than on her feet, causing her body to stretch longer and longer.

No one knows what happens beyond the event horizon because no information ever comes back out to us.

QUASARS

Quasars, first detected by radio signals in the late 1950s, look like stars but are gigantic, bright celestial objects that are really very far away. In fact, they can be 10 billion light-years away, close to the edge of the observable universe. The name *quasar* is an abbreviation for "quasi-stellar radio source." *Quasi* means "a resemblance to" and *stellar* means "star." They are called "quasi-stellar radio source" because they were first thought to be stars that were emitting radio waves. But as research advanced and telescopes became more powerful, astronomers discovered that quasars are actually active young galaxies with huge black holes at their centers. The amount of energy in quasars is hard to imagine. One quasar emits energy that is equivalent to 10 trillion suns! Astronomers think that black holes at the center of quasars swallow great amounts of matter, giving off enormous quantities of energy. That's why we can see their light. Studying quasars is essential for understanding how the universe was formed.

did you know?
....
BECAUSE QUASARS ARE BILLIONS OF LIGHT-YEARS AWAY, WHAT WE SEE TODAY THROUGH OUR TELESCOPES IS WHAT ACTUALLY HAPPENED IN THE UNIVERSE BILLIONS OF YEARS AGO.

THE BIRTH OF QUASARS ▶

Forming a quasar takes an incredible amount of mass and energy. Some quasars are formed when two or more galaxies merge. As they approach each other, their immense gravity pulls them toward each other. When they collide, a titanic explosion triggers the formation of new stars and materials. One result of two merging galaxies is that huge amounts of gases are pulled toward the central region, providing fuel for the black hole. The energy produced by the inflow of gases is so great that a quasar is formed. After this cataclysmic event, hundreds of millions of years will pass before the new quasar settles into a relatively quiet existence.

Many galaxies, including our own, have a black hole at the center.

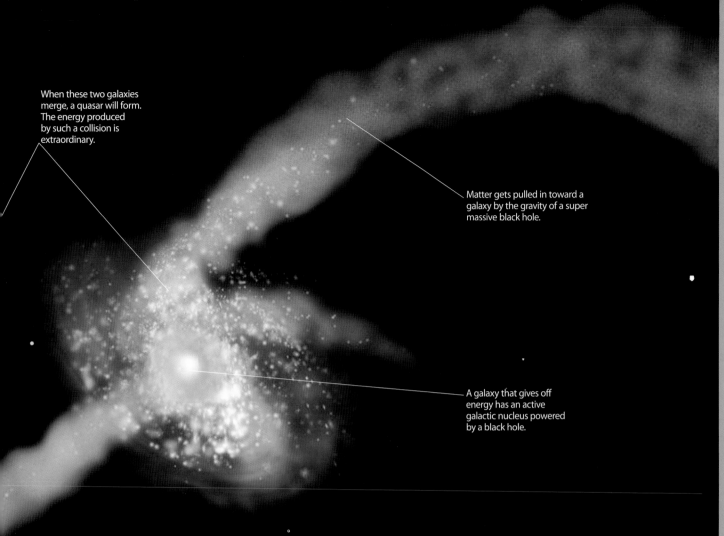

When these two galaxies merge, a quasar will form. The energy produced by such a collision is extraordinary.

Matter gets pulled in toward a galaxy by the gravity of a super massive black hole.

A galaxy that gives off energy has an active galactic nucleus powered by a black hole.

▼ EVIDENCE OF COLLISIONS

The bright quasar in the center is 5 billion light-years away from Earth. But it has no host galaxy, which is a galaxy within which a quasar is embedded. Astronomers think the quasar is the result of a collision between a normal galaxy—one that is not active the way a quasar is—and an object that had a giant black hole. The cloudlike object above is probably a disturbed galaxy—one that has undergone a recent collision. The bright star below is nowhere near the quasar.

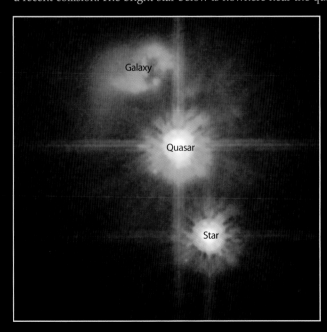

Galaxy

Quasar

Star

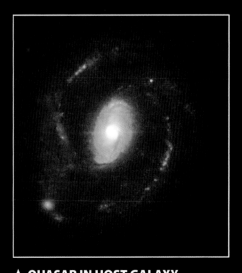

▲ QUASAR IN HOST GALAXY

A normal quasar like this one is surrounded by a host galaxy—in this case a spiral galaxy.

ASTRONAUTS

Drinking balls of floating fruit juice may be fun, but astronauts can also have it tough, especially when they suffer from "puffy-head bird-leg syndrome"! Living in microgravity—that is, almost no gravity—the fluid that is usually pulled down into the astronauts' legs stays in their face, chest, and arms. That gives them a puffy face and skinny legs, at least until the flight is done. In space, human bodies have to adjust to microgravity so that they can maintain homeostasis. Homeostasis is the condition in which the human body's internal environment is kept stable in spite of change in the outside environment. Our bodies have systems that help us stay in balance by taking in nutrients and getting rid of wastes. We breathe in oxygen and breathe out carbon dioxide. On a space shuttle or space station, maintaining homeostasis can be tricky, but training and technology have made it possible.

FOOD GOES IN ▶

Astronauts begin their meals with a pair of scissors, to cut open their airtight packages of food. Meals must, of course, contain all the nutrients that the astronauts need. But the meals also must be tidy. Food cannot be crumbly and create a mess that floats in the air—a danger to lungs and to equipment. And trash is carefully cleaned up, so there are no stray wrappers floating around.

Sweet and sour beef

Chunky chicken stew

Trail mix

Pineapple

Granola

did you
know?............................

LIQUID WASTES EJECTED INTO SPACE BECOME
CLOUDS OF TINY ICE CRYSTALS, WHICH ONE
ASTRONAUT CALLED A "BEAUTIFUL SIGHT."

Any liquid that escapes in microgravity will
form a free-floating ball shape.

Instead of drinking out of a
cup, astronauts sip beverages
from a bag with a straw.

◄ WASTE GOES OUT

When there is no gravity to help guide body wastes
to a safe storage place, vacuums and fans must do
the dirty work. After the waste is collected, getting
rid of it often means tossing it out of the space
station. It will eventually fall toward Earth and burn
up in the atmosphere.

1 The toilet seat is similar to those on Earth.

2 Bars swing over the thighs to hold the astronaut in place.

3 A vacuum sucks up solid waste and stores it in sealed bags.

4 A tube attached to a funnel collects liquid waste.

5 Air filters kill bacteria and absorb odors.

6 Foot rests can have straps that hold the astronaut's feet.

INTERNATIONAL SPACE STATION

In the early 1970s, the United States and Russia were each sending space stations into Earth's orbit. Eventually they decided to join forces to build a space station together. Eleven European countries, Canada, Brazil, and Japan joined in. The result is the International Space Station (ISS)—a laboratory orbiting about 200–250 miles (about 322–402 km) above Earth. Scientists from geologists to doctors to physicists perform experiments in the ISS, many of which have to do with the challenges of living and working in space. Cells, plants, insects, and mice have been studied in order to learn how their reproduction, growth, and health are affected by microgravity conditions.

SPECIAL DELIVERY ▼

Russian Soyuz (shown below), the U.S. Space Shuttle, and European spacecraft can dock at the ISS. They carry astronauts from all cooperating countries to and from the station. Between piloted missions, pilotless delivery vehicles, such as Russian Progress vehicles, also deliver supplies. These vehicles are computer-controlled. They may be programmed to dock with the ISS, or astronauts aboard the ISS can use a robot arm to grab a supply vehicle. While there, supply vehicles use their engine power to help keep the station in its orbit by raising its altitude and controlling its orientation. They also bring the trash back to Earth.

Canadarm2, a robotic arm to handle large objects and assist astronauts working in space

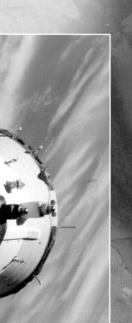

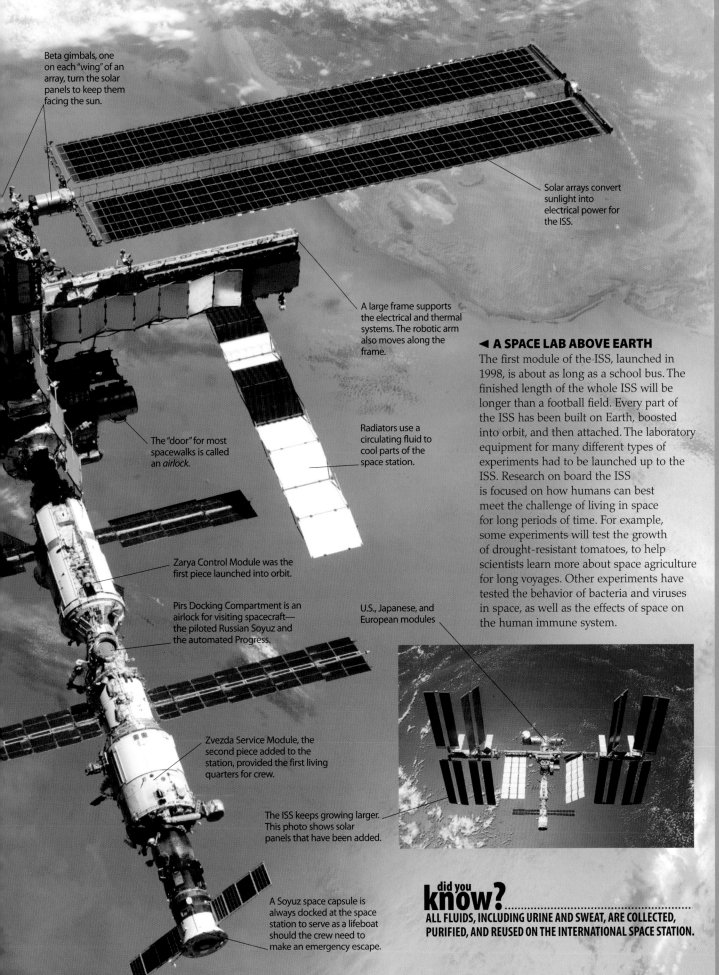

Beta gimbals, one on each "wing" of an array, turn the solar panels to keep them facing the sun.

Solar arrays convert sunlight into electrical power for the ISS.

A large frame supports the electrical and thermal systems. The robotic arm also moves along the frame.

Radiators use a circulating fluid to cool parts of the space station.

The "door" for most spacewalks is called an *airlock*.

Zarya Control Module was the first piece launched into orbit.

Pirs Docking Compartment is an airlock for visiting spacecraft—the piloted Russian Soyuz and the automated Progress.

U.S., Japanese, and European modules

Zvezda Service Module, the second piece added to the station, provided the first living quarters for crew.

The ISS keeps growing larger. This photo shows solar panels that have been added.

A Soyuz space capsule is always docked at the space station to serve as a lifeboat should the crew need to make an emergency escape.

◄ A SPACE LAB ABOVE EARTH

The first module of the ISS, launched in 1998, is about as long as a school bus. The finished length of the whole ISS will be longer than a football field. Every part of the ISS has been built on Earth, boosted into orbit, and then attached. The laboratory equipment for many different types of experiments had to be launched up to the ISS. Research on board the ISS is focused on how humans can best meet the challenge of living in space for long periods of time. For example, some experiments will test the growth of drought-resistant tomatoes, to help scientists learn more about space agriculture for long voyages. Other experiments have tested the behavior of bacteria and viruses in space, as well as the effects of space on the human immune system.

did you know? ....
ALL FLUIDS, INCLUDING URINE AND SWEAT, ARE COLLECTED, PURIFIED, AND REUSED ON THE INTERNATIONAL SPACE STATION.

GLOSSARY

alloys

Two metal elements combined to produce a stronger whole. Alloys can have different properties depending on which two elements are mixed together.

antigens

Chemical molecules within red blood cells that determine blood types. There are four different blood types: A, B, AB, or O.

arthropods

Arthropods are bilaterally symmetrical and have an exoskeleton. They include insects and arachnids (spiders).

asteroid

Lumps of rock and metal clustered together in space. The asteroid belt (high concentration of asteroids) can be found between Mars and Jupiter.

atmosphere

Layers of gas (decreasing in density the further they are from Earth) that surround and protect the planet. The different layers are called the troposphere, stratosphere, mesosphere, thermosphere, and exosphere.

aurora

A colored glow in the sky resulting from a reaction between cosmic rays (charged particles deflected by the Earth) and atoms in the atmosphere.

bacteria

Single-celled organisms that grow and survive in almost every habitat on Earth. Bacteria are called 'extremophiles' because of their ability to survive in extreme conditions.

Big Bang theory

The scientific theory of how the galaxies were formed. The universe was once a small, focused point but after a huge burst of energy began to expand, spreading matter across space.

bilaterally symmetrical

Organisms where the left and right sides of the body are mirror images of one another. A spider or butterfly is bilaterally symmetrical.

biodiversity

The various different life forms that exist within a biome or ecosystem.

biome

A category for a huge zone on Earth. The rainforest is one biome and the desert is another. Ecosystems exist within biomes.

brainstem

The stem joining the brain and spinal cord. It relays messages, from nerves running through the body and up the spinal cord, into the brain.

celestial body

Any naturally occurring object in space. The planets, asteroids, and meteorites are all celestial bodies.

Cenozoic

A term to describe the period of history spanning the last 65 million years. The Paleozoic and the Mesozoic eras precede the Cenozoic.

chemosynthesis

The process by which deep-sea organisms convert chemicals from hydrothermal vents on the seabed into energy, as opposed to photosynthesis when sunlight is converted into energy.

digestive system

The organs in the body, such as the intestine and stomach, that break down food, process nutrients, and get rid of any waste products.

DNA (deoxyribonucleic acid)
A molecule, existing in the nucleus of a cell, that contains all the information that an organism needs to grow and develop.

ectotherms
Organisms that are unable to regulate their own body temperature. Unlike humans whose body temperatures are constant, ectotherms must adapt their behavior to alter the amount of heat they lose or gain.

enzymes
Molecules that speed up chemical reactions within an organism; for example, enzymes have been genetically removed from crops to slow down the rate at which they rot.

equator
The imagined line dividing the Northern and Southern hemispheres.

exoskeleton
The hard casing, or shell, that protects an animal's soft tissue from predators and regulates water loss.

fetus
An embryo develops into a fetus after 8 weeks inside a woman's uterus. The fetus will develop and after approximately 40 weeks will become a baby with the ability to survive outside the womb.

galaxy
A group of many stars. The sun is a star in the Milky Way Galaxy. There are millions of galaxies in the universe.

geothermal energy
Heat from within the Earth. Heated water from underground can be converted into energy above ground. Geysers are natural eruptions of geothermal water and steam.

gravity
A force that pulls two objects towards each other. The higher the mass of an object, the stronger the gravitational pull will be. Microgravity occurs when the gravitational pull is less strong.

greenhouse gases
The name given to gases such as carbon dioxide and methane that trap heat within the Earth's atmosphere. An increase in greenhouse gases can lead to global warming.

homeostasis
The process by which a human's internal environment remains stable when subjected to varying external environments. The human body achieves this in a number of ways, including getting rid of waste products and regulating body temperature.

hypothalamus
A cluster of cells that regulate body functions associated with the nervous and endocrine systems; these include breathing, the release of the growth hormone, and sleep cycles.

immunoglobulin E (IgE)
Antibodies produced by white blood cells to combat allergen attacks. Immunoglobulin joins to mast cells and triggers the production of white blood cells when allergens are present.

ions
Oppositely charged particles that bond atoms together. Ions are formed when there's a change in the number of electrons in any substance.

kidneys
Organs in the body that remove waste and control water and salt levels. If the kidneys fail a person can have dialysis or a transplant.

kinetic energy
The energy of an object when it is in motion. Inertia is a resistance to a change in motion and potential energy is the energy of a stationary object in relation to its position and size.

lava
Formed when molten rock, called magma, reaches the Earth's surface. Lava erupts either in the form of rock and ash or as a thick substance that flows along the ground.

leptons and quarks
The tiniest pieces of matter that scientists believe make up the universe. The building blocks of atoms are either leptons or quarks. An electron is a type of lepton and a proton is a type of quark.

mammal
A group of animals that have hair and that usually give birth to live offspring. Mothers feed their young from mammary glands on their bodies.

melanin
The pigment that determines certain characteristics such as skin and eye color. Melanin protects the skin from ultraviolet rays from the sun.

meteorite
A fragment of rock or metal that has broken away from a larger celestial body, often when two asteroids collide. Meteorites fall to Earth, attracted by the Earth's gravitational pull.

molecule
Small particles made up of a group of atoms held together by strong chemical bonds. A group of just two atoms can make up a molecule.

momentum
Every moving object has momentum. It is the calculation of the mass of an object times its velocity (its speed with direction).

nuclear fusion
Nuclear reactions change atoms from one element to another. Nuclear fusion takes place when two atoms join to make a heavier element. A by-product of nuclear fusion is energy.

photosynthesis
The process by which plants convert energy from the sun, combining it with carbon dioxide and water, into food. Oxygen is a by-product of photosynthesis.

pollination
A means of reproduction in plants where pollen is transferred from one plant to another. Pollen can be transferred on the wind or by animals called pollinators, such as bees or hummingbirds.

quasar
A naturally occurring object in space that can emit radio waves and other types of energy.

radioactive decay
The central part of an atom holding most of the genetic material is called the nucleus. Nuclei can emit particles in a process called radioactive decay, causing an atom to change type.

refraction
Wavelengths travel through one medium to another at different speeds. Light wavelengths travel more slowly through water than air, causing light hitting water to be refracted at a slightly different angle.

respiration
The chemical reactions that take place when oxygen enters cells in the body and is converted into energy for the cells to use.

retrograde rotation
An object that spins around in the opposite direction from its trajectory. For example, Venus rotates (spins around) in the opposite direction from which it orbits the sun.

saltation
The process by which sand-sized particles are picked up, and then dropped, by the wind, forming dunes.

sedimentary rock
Sediments such as sand and mud settle on river and seabeds. More layers settle on top and they are compounded over time to form sedimentary rock.

solar system
Part of the universe made up of the sun, eight planets (including Earth), dwarf planets, moons, comets, and asteroids. The gravity of the sun keeps the planets in orbit.

sonic boom
The result of an object traveling through air faster than the speed of sound. The object causes a sudden increase in pressure that forces molecules in the air to bunch together, causing a sonic boom.

source region

An area where an air mass and the land or water below share similar characteristics. For example, a cold air mass over a polar region. When two air masses meet it is called a weather front.

specialization

The process by which stem cells change into different types of cell within the body, specialized to perform a particular function.

stalagmite

Formations of minerals extending from cave ceilings. Water from a stalagmite can drip to the cave-floor, forming stalactites that grow upwards.

stem cells

Embryonic stem cells differentiate (change into) different types of specialized cells. Adult stem cells change and divide to replace old cells.

supernova

A star that is dying (running out of hydrogen) swells to form red giant or supergiant stars which eventually explode. The exploding star is called a supernova.

tectonic plates

The blocks of rock that make up the Earth's surface. The edges of the plates are called faults. Pressure builds along these faults and this can cause earthquakes.

thermal energy

Energy created from heat. Thermal energy can be transferred by conduction, convection, or by radiation via electromagnetic waves.

tsunami

A surge of seawater, resulting in a massive rolling wave, caused by eruptions between Earth plates on the sea floor.

tumor

Caused by the rapid growth and division of cancer cells. These abnormal growths can be surgically removed or treated with chemotherapy or radiation therapy.

universe

All existing matter within space. The Earth and solar system are all part of the universe.

vaccine

Contains weakened virus antigens that stimulate active immunity (the production of antibodies to kill a disease) ensuring the body will be able to fight off any further attacks.

virus

Chemical packages (smaller than bacteria) that reproduce their genetic material by invading other cells. They cause illnesses such as smallpox and the common cold.

voltage

The force exerted to charge an electrical circuit, measured in volts.

wavelength

The distance between one wave peak, or trough, and the next. Radio, sound, and light waves all have different wavelengths, and can change speed from one medium to another.

INDEX